Diagnosis of Onychomycosis and Other Nail Disorders

A PICTORIAL ATLAS

Springer
New York
Berlin
Heidelberg
Barcelona
Budapest
Hong Kong
London
Milan
Paris
Santa Clara
Singapore
Tokyo

Diagnosis of Onychomycosis and Other Nail Disorders

A PICTORIAL ATLAS

C. Ralph Daniel III, MD

CLINICAL PROFESSOR OF MEDICINE (DERMATOLOGY)
University of Mississippi Medical Center
Jackson, Mississippi

Photographic Contributions by:
Gary D. Palmer, MD
C. Ralph Daniel III, MD
Phoebe Rich, MD
Boni E. Elewski, MD
Raza Aly, PhD
Rosemary T. DePaoli, MD
American College of Rheumatology
Syntex

With 60 Illustrations

Springer

C. Ralph Daniel III, MD
Clinical Professor of Medicine (Dermatology)
University of Mississippi Medical Center
Jackson, MS 39216 USA

Library of Congress Cataloging-in-Publication Data applied for.

Printed on acid-free paper.

©1996 Springer-Verlag New York, Inc.
All rights reserved. This work may not be translated or copied in whole or in part without the written permission of the publisher (Springer-Verlag New York, Inc., 175 Fifth Avenue, New York, NY 10010, USA), except for brief excerpts in connection with reviews or scholarly analysis. Use in connection with any form of information storage and retrieval, electronic adaptation, computer software, or by similar or dissimilar methodology now known or hereafter developed is forbidden.

The use of general descriptive names, trade names, trademarks, etc., in this publication, even if the former are not especially identified, is not to be taken as a sign that such names, as understood by the Trade Marks and Merchandise Marks Act, may accordingly be used freely by anyone.

While the advice and information in this book are believed to be true and accurate at the date of going to press, neither the authors nor the editors nor the publisher can accept any legal responsibility for any errors or omissions that may be made. The publisher makes no warranty, express or implied, with respect to the material contained herein.

Printed and bound by Walsworth Publishing Co., Marceline, MO.
Printed in the United States of America.

9 8 7 6 5 4 3 2 1

ISBN 0-387-94625-X Springer-Verlag New York Berlin Heidelberg

CONTENTS

I. **INTRODUCTION • 1**

II. **INFLAMMATORY NAIL DISORDERS • 5**

　A. **Psoriasis and Psoriasis-like Nail Disorders • 6**

　　1. Psoriasis • 6
　　2. Pityriasis rubra pilaris • 17
　　3. Reiter's Syndrome • 21

　B. **Other Inflammatory Nail Disorders • 24**

　　1. Lichen planus • 24
　　2. Darier's disease • 34
　　3. Idiopathic twenty-nail dystrophy • 38
　　4. Eczematous dermatitis • 40
　　5. Simple onycholysis • 42

III. **TRAUMATIC NAIL DISORDERS • 47**

　A. **Onychogryphosis • 48**

　B. **Habit tic nail disorders • 51**

　C. **Transverse striate leukonychia • 52**

　　1. Traumatic • 52
　　2. Nontraumatic (Mee's lines) • 53

　D. **Leukonychia punctata • 55**

IV. **LOCAL INFECTIONS • 57**

　A. **Onychomycosis • 58**

　　1. Distal subungual onychomycosis • 58
　　2. Proximal subungual onychomycosis • 62
　　3. White superficial onychomycosis (leukonychia mycotica) • 64
　　4. Total dystrophic onychomycosis • 66
　　5. *Candida* onychomycosis • 66

　B. **Paronychia • 66**

　　1. Acute paronychia • 68
　　2. Chronic paronychia • 71

　C. **Other infectious conditions affecting the nails • 72**

　　1. Chronic mucocutaneous candidiasis • 72
　　2. Nail changes in patients with HIV infection • 74
　　3. Mycotic keratoderma • 76
　　4. Green Nail • 78
　　5. Crusted scabies • 81

GLOSSARY • 82

REFERENCES • 84

APPENDIX A • 86

APPENDIX B • 88

INDEX • 89

I wish to dedicate this book to my family;

Melissa, Carl, and Jon Daniel.

Without their support and

understanding, I would never

have had the time for this endeavor.

CHAPTER ONE

INTRODUCTION

Nails are subject to alterations by both exogenous and endogenous forces, including numerous environmental factors (e.g., moisture, irritants, trauma, nail cosmetics) as well as systemic diseases. There are also a number of anatomic and physiologic limitations of the nail plate, thereby rendering diagnosis a challenge, even for the most experienced clinician. Figure 1 depicts a diagrammatic representation of the basic components of the nail itself, which will be referred to throughout this book.

Figure 2 shows normal nails and will provide a dramatic counterpoint to the many "abnormal" nails which are displayed and discussed. A nail examination should be conducted in adequate, preferably natural, light and should include all nails.[1] All nail cosmetics should be removed from the nail, and it may be helpful to cleanse the nail with a solvent such as alcohol or acetone. Unfortunately, there are few hard and fast rules that dictate the steps to be followed in a nail examination. In some cases, local factors (e.g., fungal infection, trauma, a tumor, circulatory problems) affect only a few nails. In addition, certain disorders (e.g., nail tumors, systemic diseases, contactants) are more common to specific areas of the nail, such as the nail bed, the nail matrix or the nail plate, than others, while other disorders (e.g., psoriasis, lichen planus) may affect any or all portions of the nail unit. In some cases, physical examination of the nail is all that is required to arrive at a diagnosis. More commonly, however, additional information is needed, including a thorough history and various laboratory tests.

Appendix A includes a screening questionnaire, which may provide useful information in arriving at a clinical diagnosis. Some of the most widely utilized laboratory tests used to diagnose nail disorders include a potassium hydroxide (KOH) preparation, fungal and bacterial cultures, nail biopsy with appropriate stains, and microscopy.

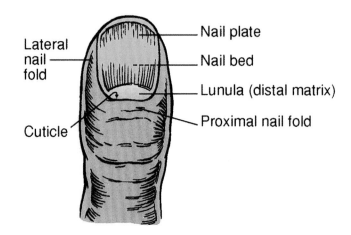

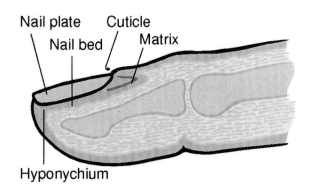

Figure 1. Diagrammatic representation of the normal nail structure

In general, nail disorders may be classified according to whether they are hereditary, congenital, or acquired, although in some cases there is not a clear delineation between categories. Table 1 provides a very broad classification that may be used for the initial categorization of a nail disorder, prior to completing a thorough investigation. The text and illustrations that follow provide many clinical and laboratory clues for differentiating between some of the most common nail disorders, including onychomycosis and other conditions that mimic it. It is hoped that these will assist both the dermatologist and the general practitioner in arriving at a differential diagnosis.

INTRODUCTION

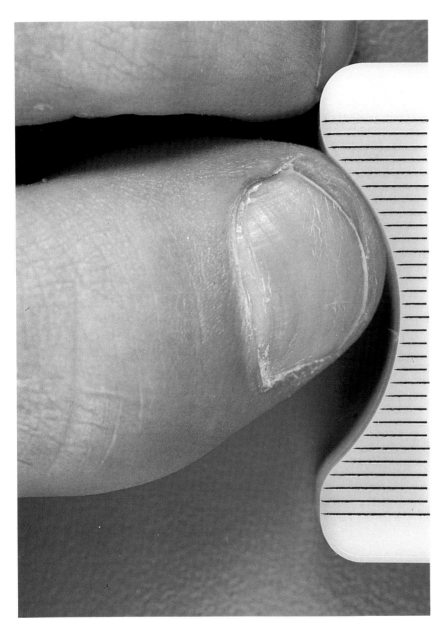

Figure 2. Normal nail *(Courtesy of Raza Aly, MD)*

TABLE 1	General classification schema for nail examination[1]

- Primarily dermatologic condition
- Primarily systemic disorder
- Systemic drug(s)
- Bacterial, fungal or viral organism(s)
- Local or topical agent(s) in the vicinity of the nail
- Benign or malignant tumor(s)
- Physical agents (trauma caused by shoes, "corn plasters," etc.)
- Other

CHAPTER TWO

INFLAMMATORY NAIL DISORDERS

II.A. PSORIASIS AND PSORIASIS-LIKE NAIL DISORDERS

A.1. PSORIASIS

Psoriasis is a condition with a strong genetic component. While only about 2% to 3% of Americans suffer from psoriasis, approximately one out of every three people who come from families with a history of psoriasis will have this condition. Nail involvement occurs in approximately 10% to 50% of patients with psoriasis. However, most patients with psoriasis of the nails have psoriasis elsewhere on their bodies, including the elbows, knees, scalp and lumbosacral area, and fewer than 10% of patients present with psoriasis of the nails only.[2] As shown in Table 2, the clinical signs of psoriasis of the nails depend on the portion of the nail unit that is involved.

A.1.a. Clinical Presentation

One of the key signs of psoriasis of the nails is pitting, which results from the development of small psoriatic lesions in the proximal portion of the nail matrix (Figure 3). The pits are usually large, relatively deep, and scattered. In some cases, deeper lesions may occur in the more distal portion of the nail matrix. Another characteristic sign of psoriasis of the nails is a salmon patch or lesion known as "the oil drop change" (Figure 4). The oil drop change is caused by the exudation of a serum glycoprotein, which "weeps" onto the nail bed. However, it should be noted that the oil drop change may also occur in patients with systemic lupus erythematosus.[3]

Psoriasis of the nail bed may also result in mild to severe subungual hyperkeratosis (Figure 5). In some cases, extensive cornification may cause uplifting of the nail plate. It may also cause onycholysis, which results from a split in the intracornified cell layer, resulting in the separation of the nail plate from its underlying anchorage (Figure 6). Another common finding in psoriasis of the nails is splinter hemorrhages, surface pinpoint bleeding points which appear in the nail bed (Figure 7).

Other, less common features of psoriasis of the nails include onychorrhexis (see "Lichen Planus") and Beau's lines (Figure 8). Onychorrhexis is associ-

TABLE 2	Clinical Signs of Psoriasis[2]
Anatomic Site	**Clinical Sign**
Distal matrix	Thinned nail plate, erythema of the lunula
Hyponychium	Subungual hyperkeratosis, onycholysis
Intermediate matrix	Leukonychia
Nail Bed	"Oil drop" sign or "salmon patch", subungual hyperkeratosis, onycholysis, splinter hemorrhages
Nail plate	Crumbling and destruction plus other changes secondary to the specific site
Phalanx	Psoriatic arthritis with nail changes over 80 per cent of the time
Proximal and lateral nail folds	Cutaneous psoriasis
Proximal matrix	Beau's lines, pitting, onychorrhexis

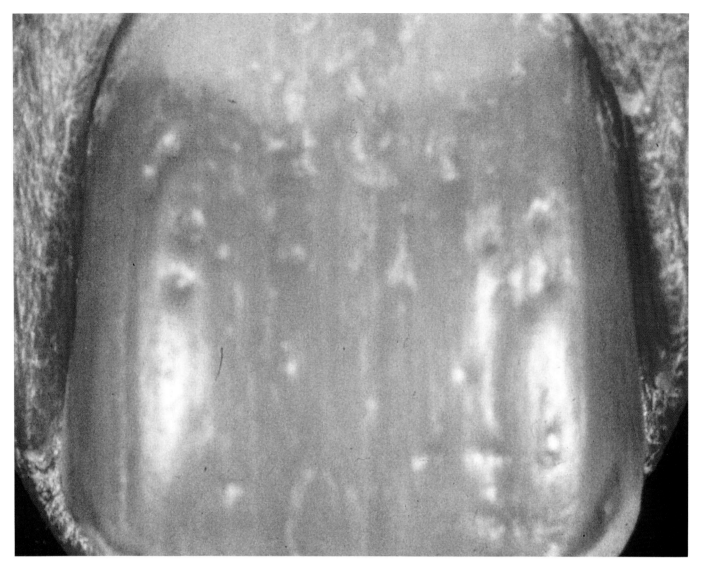

Figure 3. Pitting of the nail plate *(Courtesy of American College of Rheumatology)*

ated with irregular, longitudinal ridging of the nail plate, while Beau's lines appear as deep single or multiple transverse ridges. Beau's lines are due to intermittent inflammation of the proximal nail fold region, with resultant injury to the underlying proximal nail matrix. Leukonychia of the nail plate due to psoriasis (Figure 9) may occur when the intermediate portion of the nail matrix is involved in the psoriatic process. The extent of this condition is correlated with the amount of psoriasis in this area.

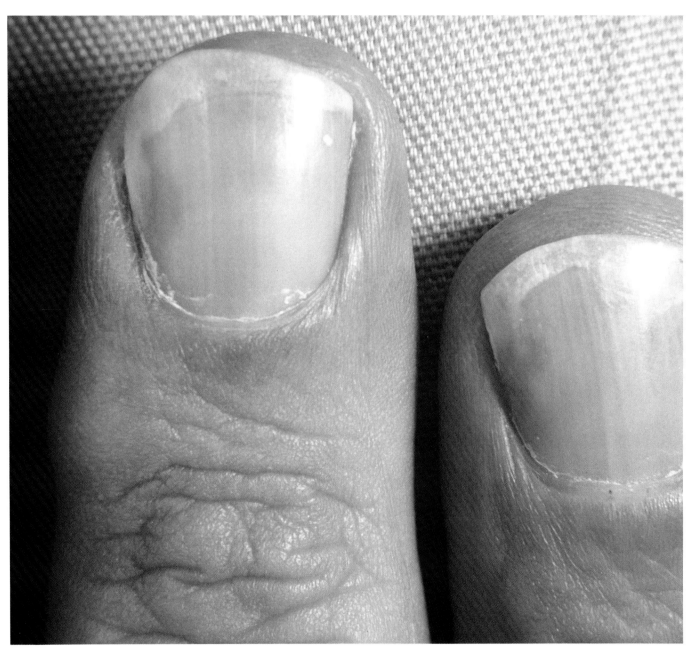

Figure 4. "Oil drop" change or salmon patch due to psoriasis *(Courtesy of C.R. Daniel III, MD)*

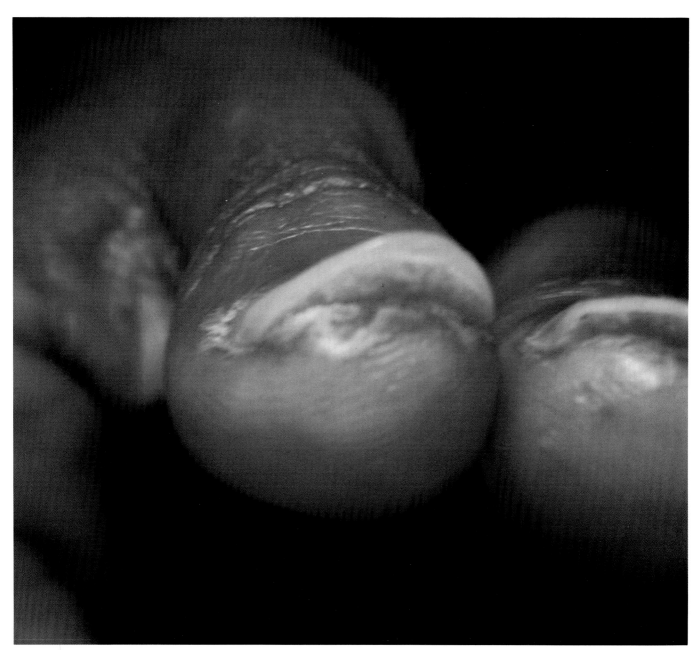

Figure 5. Subungual hyperkeratosis *(Courtesy of C.R. Daniel III, MD)*

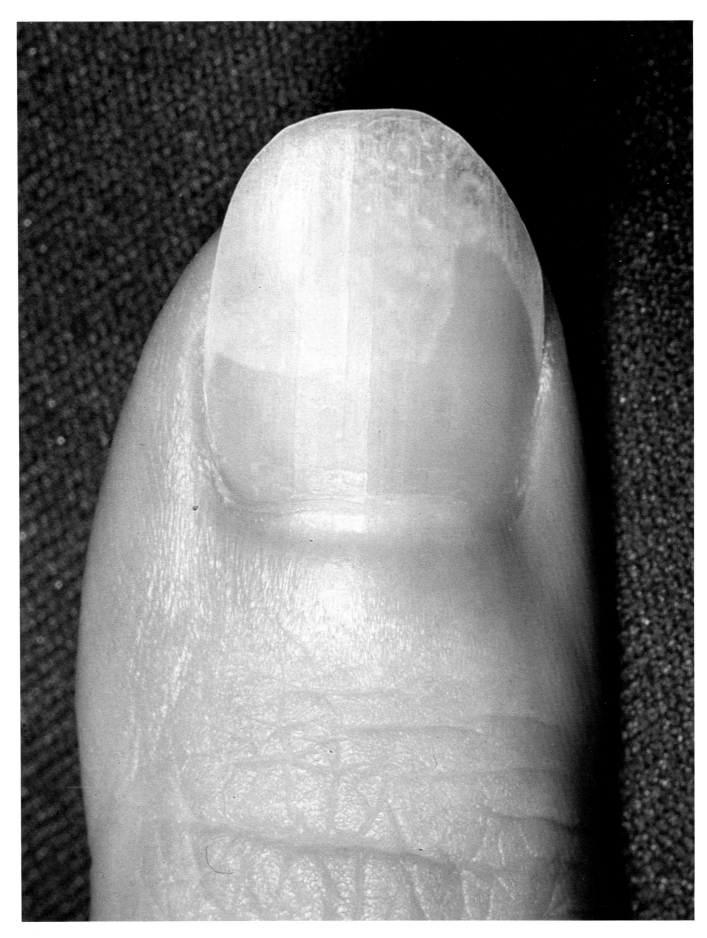

Figure 6. Distal onycholysis due to nail bed psoriasis *(Courtesy of C.R. Daniel III, MD)*

INFLAMMATORY NAIL DISORDERS

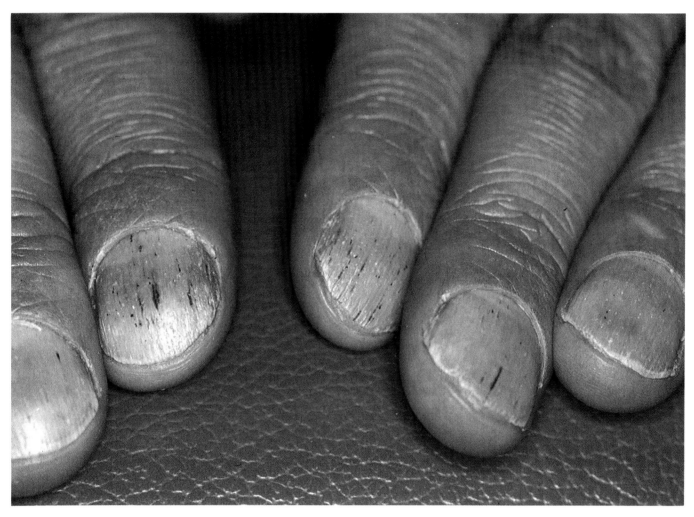

Figure 7. Splinter hemorrhages *(Courtesy of Phoebe Rich, MD)*

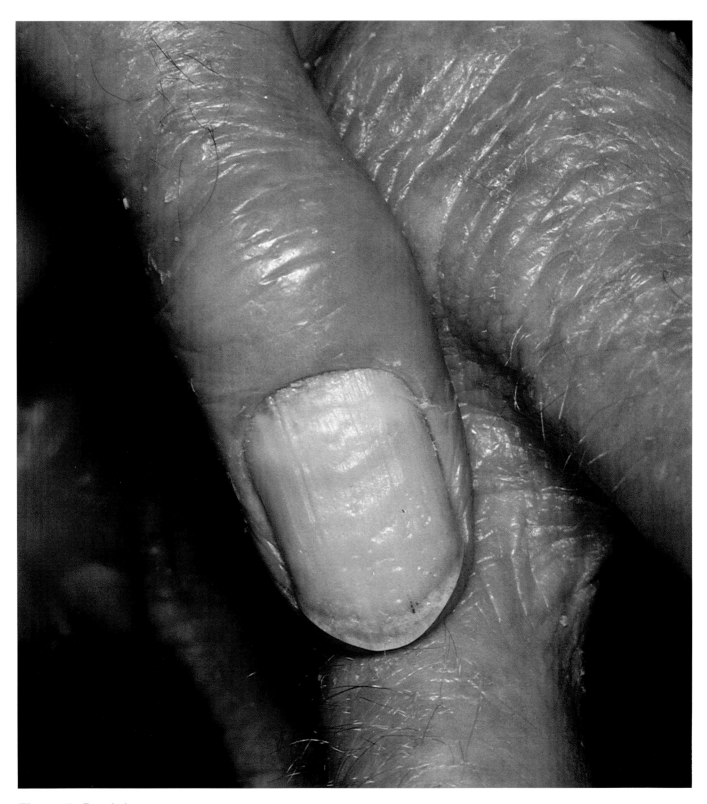

Figure 8. Beau's lines *(Courtesy of Gary Palmer, MD)*

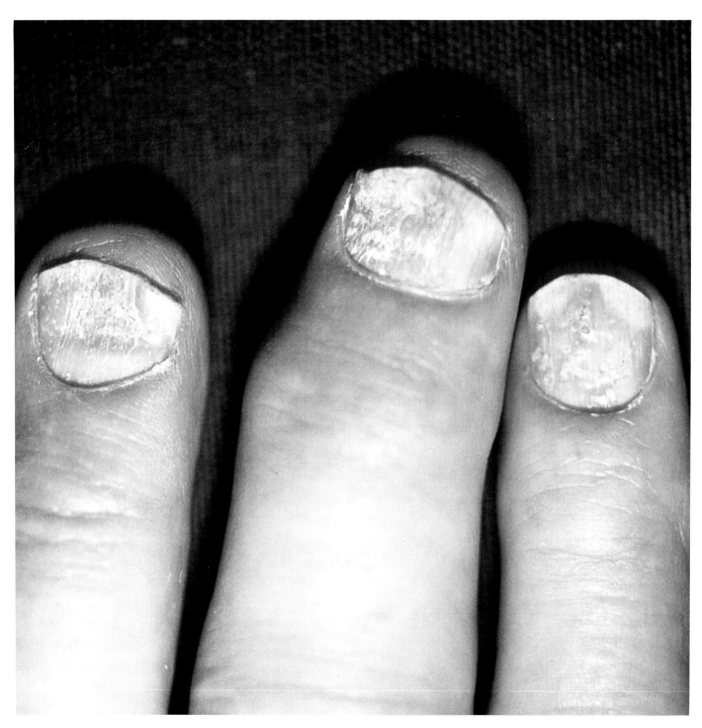

Figure 9. Leukonychia of nail plate due to psoriasis *(Courtesy of American College of Rheumatology)*

A.1.b. Patient History

Patients with psoriasis may typically develop a psoriatic lesion whenever they traumatize a nail or any area on the skin. Therefore, a careful history is important in determining whether the patient traumatizes certain nails more often than others. If so, he or she is more likely to develop psoriasis in those particular nails.

A.1.c. Differential Diagnosis

As previously noted, the vast majority of patients with psoriasis have visible evidence of active psoriasis. Thus, it is easier to diagnose psoriasis of the nails when one or more of the nail changes described above occur concomitantly with the cutaneous manifestations of this condition. However, diagnosis is more challenging in the small percentage of patients who present with psoriasis of the nails alone. In such cases, it may be much more difficult clinically to distinguish psoriasis of the nails from onychomycosis. The following recommendations may be used for achieving a differential diagnosis.

Clinically, an important clue that the nail condition is psoriasis, and not onychomycosis, is the fact that psoriasis typically involves both hands, whereas onychomycosis in the nonimmunocompromised host is usually limited to one hand. However, both conditions may involve the feet as well. Mycologic studies should be performed to rule out a diagnosis of onychomycosis, including a KOH preparation and suitable fungal cultures. In psoriatic patients, fungal cultures are generally negative for fungi, although a positive culture may be obtained from the toenails of psoriatic patients as a result of concomitant, secondary infection. Psoriasis and dermatophyte infections even more rarely co-exist on fingernails.

A nail biopsy and histologic examination should be considered if the KOH preparations and cultures are negative, and if the diagnosis is still in doubt. The tissue should be stained with periodic acid-Schiff (PAS), as well as hematoxylin and eosin. Table 3 lists the histopathologic changes associated with nail psoriasis; these are illustrated in Figures 10 through 13.

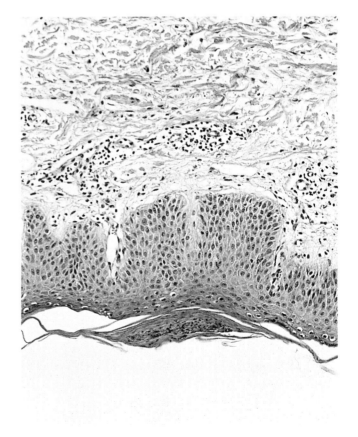

Figures 10–13. Histophathologic representations of psoriasis. (*Courtesy of C.R. Daniel III, MD*)

| TABLE 3 | Clinical signs of psoriasis of the nail[4] |

- Hemmorhage in cornified layer
- Hypergranulosis
- Marked hyperkeratosis
- Papillomatous epidermal hyperplasia
- Serum globules in cornified layer
- Spongiosis

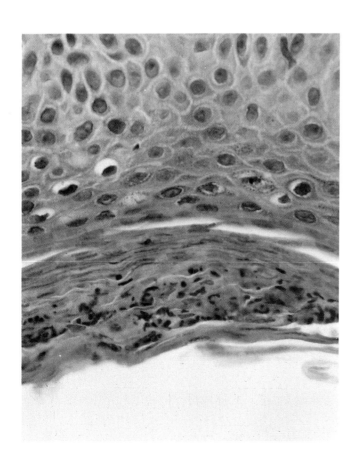

In addition to the similarities between onychomycosis and psoriasis, parakeratosis pustulosa is (Figure 14) a condition that resembles psoriasis and usually develops in girls under the age of five. One of the clues that can be used to differentiate these two disorders in children of this age is that parakeratosis pustulosa almost always appears on just one nail — usually a fingernail. The nail may be tender and swollen; it may also be brittle and enlarged. Fortunately, this condition almost always resolves spontaneously.

Figure 14. Parakeratosis pustulosa
(*Courtesy of C.R. Daniel III, MD*)

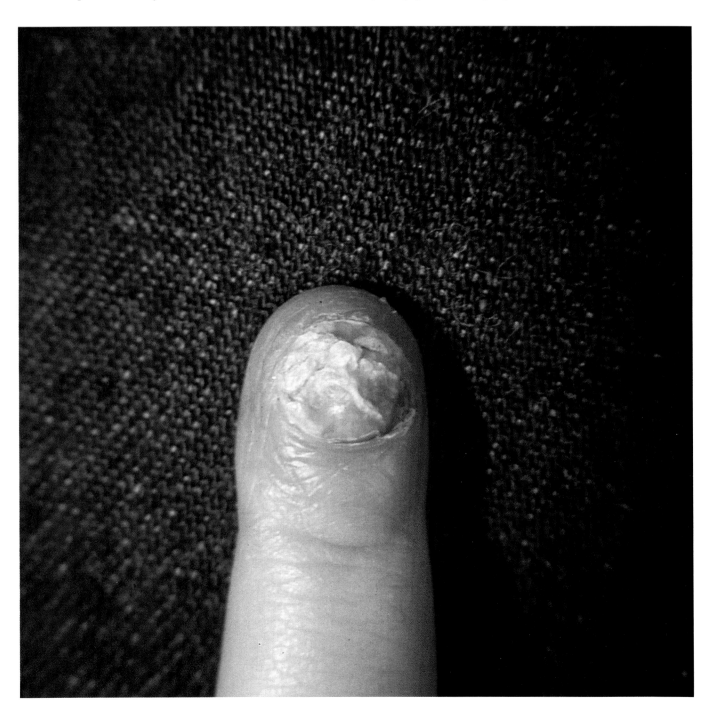

II.A.2. PITYRIASIS RUBRA PILARIS

A.2.a. Clinical Presentation

Pityriasis rubra pilaris is a cutaneous disorder characterized by follicular acuminate papules (Figure 15). Many patients with this condition have nail involvement, which in some cases may be mistaken for psoriasis of the nails. Pityriasis rubra pilaris of the nails may be suggested by yellow-brown discoloration of the nail plate (Figure 16), subungual hyperkeratosis (nail thickening) (Figure 17), and cutaneous involvement of the paronychium. Splinter hemorrhages may also be found. In addition, patients with pityriasis rubra pilaris of the nails frequently have thickening of the skin on the palms of their hands or on the soles of their feet. Extensive pityriasis rubra pilaris-type skin changes are also common and consist of scattered normal areas in the middle of the involved area known as "skip" areas (Figure 18).

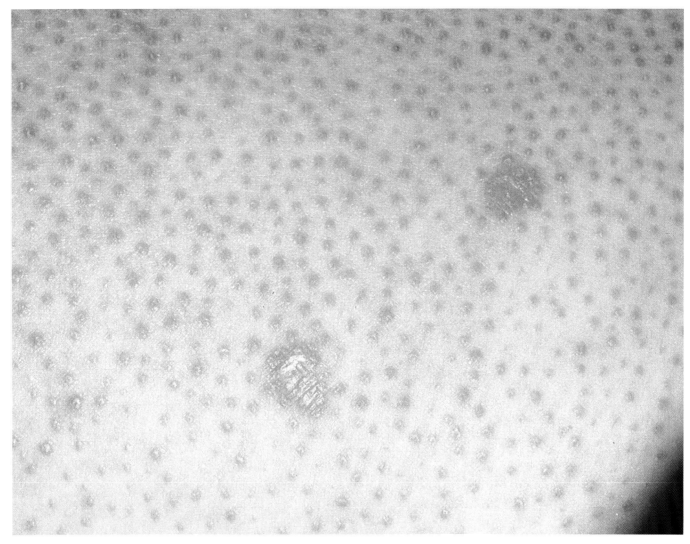

Figure 15. Follicular acuminate papules suggestive of rubra pilaris *(Courtesy of Gary Palmer, MD)*

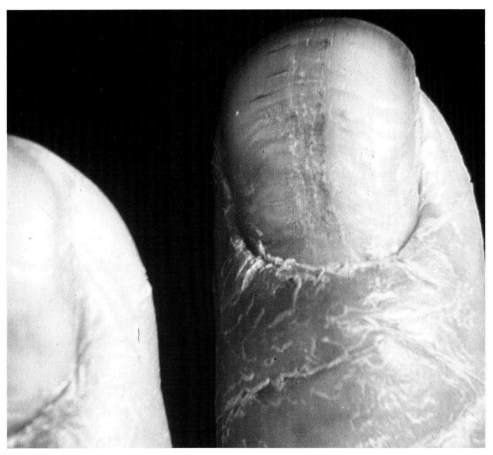

Figure 16. Yellow-brown discoloration of the plate associated with pityriasis rubra pilaris *(Courtesy of Gary Palmer, MD)*

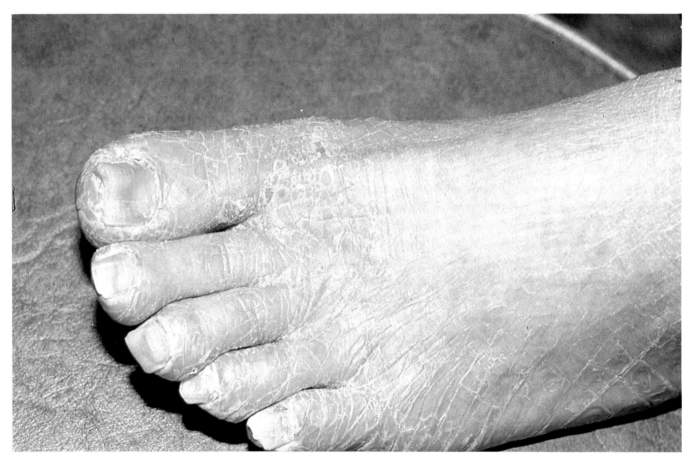

Figure 17. Subungual hyperkeratosis associated with pityriasis rubra pilaris *(Courtesy of Gary Palmer, MD)*

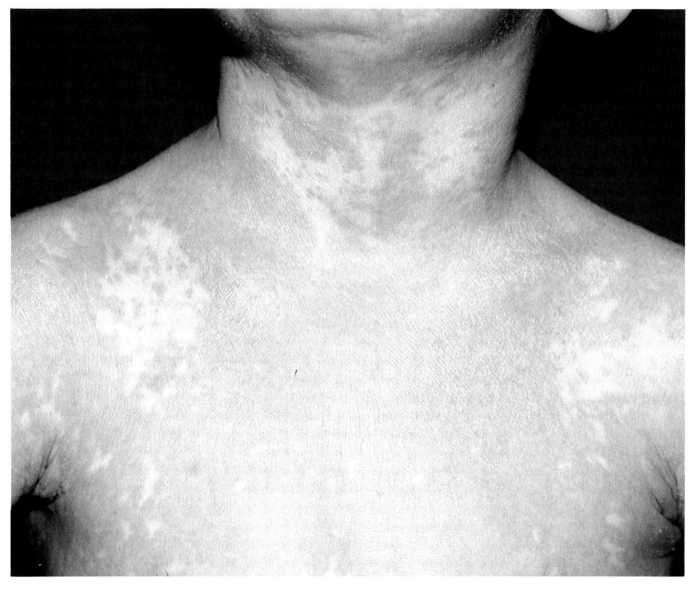

Figure 18. "Skip" area on the trunk of a patient with pityriasis rubra pilaris *(Courtesy of Gary Palmer, MD)*

A.2.b. Differential Diagnosis

Onycholysis is less frequent in pityriasis rubra pilaris than in psoriasis of the nails. There is also significantly less pitting, fewer indentations and less chance of seeing an oil drop change. The cutaneous "skip" lesion characteristic of pityriasis rubra pilaris does not occur in patients with psoriasis. Other characteristic findings that may distinguish pityriasis rubra pilaris from psoriasis include the presence of acuminate papules around the hair follicles or a peri-follicular distribution on the body. As is the case with psoriasis of the nails, fungal cultures are usually negative.

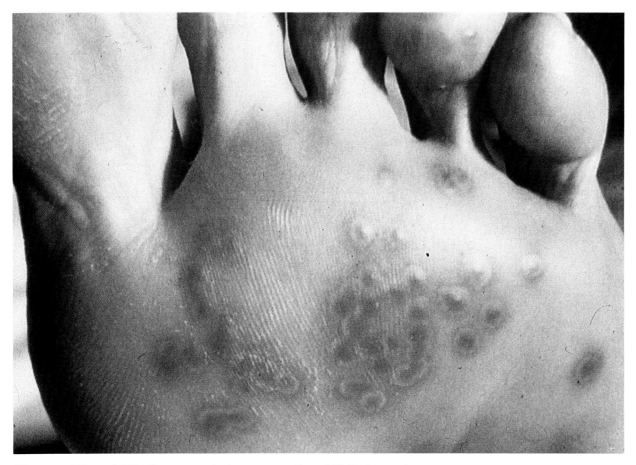

Figures 19 and 20. Cutaneous lesions suggestive of Reiter's syndrome
(*Courtesy of American College of Rheumatology*)

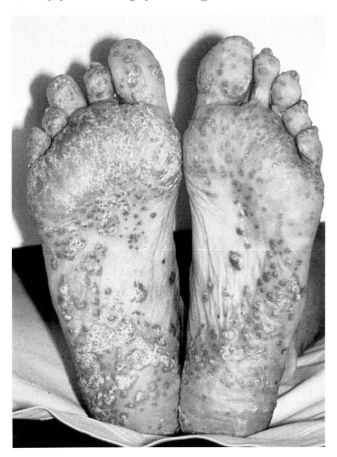

II.A.3. REITER'S SYNDROME

A.3.a. Clinical Presentation

Reiter's syndrome is a constellation of diseases consisting of arthritis, conjunctivitis, urethritis and cutaneous lesions, which may be inherited by an HLA-B27 pattern. Both the cutaneous manifestations (Figures 19 and 20) and the histologic changes of Reiter's syndrome may be quite similar to psoriasis. Patients with Reiter's syndrome may develop nail changes virtually identical to psoriasis of the nails, although this is a much less common occurrence. The primary nail changes associated with Reiter's disease are onycholysis, yellowing of the nail plate, and subungual hyperkeratosis (Figures 21 and 22). In addition, small yellow peri- and subungual pustules may develop beneath the nail plate, often near the lunula.[5] These may enlarge and lead to erosion through the nail plate. Paronychia-like changes may also be noted.

Figure 21. Reiter's syndrome - early *(Courtesy of Syntex)*

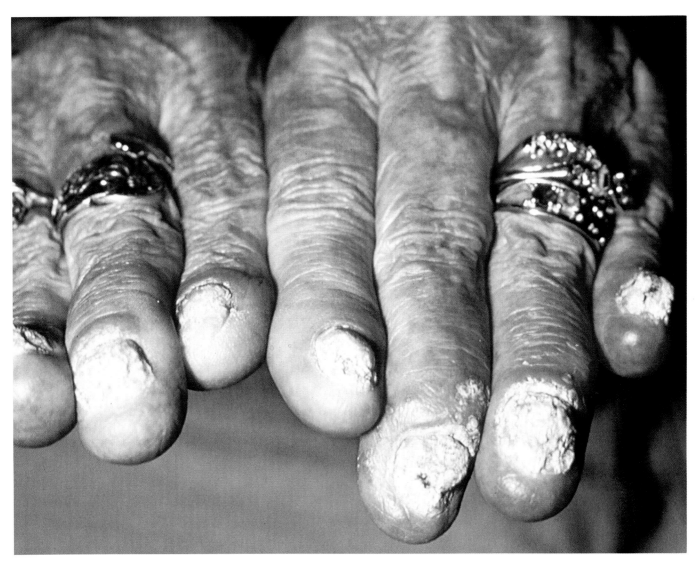

Figure 22. Reiter's syndrome - late *(Courtesy of Syntex)*

A.3.b. Differential Diagnosis

Certain factors may be used to differentiate Reiter's syndrome from psoriasis. In general, pitting is not as extensive in Reiter's disease. Additionally, patients with Reiter's syndrome have both urethritis as well as arthritis, while patients with psoriasis are relatively unlikely to have arthritis.

II.B. OTHER INFLAMMATORY NAIL DISORDERS

II.B.1. LICHEN PLANUS

B.1.a. Clinical Presentation

Lichen planus is an inflammatory skin disease characterized by wide, flat, violaceous and itchy papules, which may have a violaceous appearance (Figure 23). It may occur in circumscribed patches and is often very persistent. Nail involvement may occur in about 10% of patients with lichen planus and may appear in the absence of any involvement of the skin or mucous membranes. In common with psoriasis, lichen planus may affect the fingernails on both hands and the toenails on both feet. It may also affect the mucous membranes of the mouth, the penis, and the vagina. The nail changes typical of lichen planus are not pathognomonic but may be highly suggestive for the disease and, as is the case with psoriasis, are dependent on the location of the affected nail unit.

The most common nail manifestation of lichen planus is onychorrhexis, otherwise known as "exaggerated longitudinal ridging" (Figure 24). This may result in splitting of the nail when the entire length of the nail matrix is involved. Another characteristic finding in lichen planus is a phenomenon known as the "angelwing deformity," which occurs when diffuse atrophy of the nail matrix leads to functional shortening, causing thinning of the nail plate on each side (Figure 25). This change may be permanent. Another permanent nail change associated with lichen planus is pterygium, which occurs when severe destruction of the nail matrix results in the formation of a scar, causing a break in the nail plate connecting the proximal nail and nail bed epithelium (Figure 26). Anonychia occurs when there is total destruction of the nail matrix (Figure 27). Lichen planus of the nails may also be associated with yellow discoloration, onycholysis, hyperpigmentation, and subungual hyperkeratosis.

Figure 23. Cutaneous manifestations of lichen planus
(Courtesy of Gary Palmer, MD)

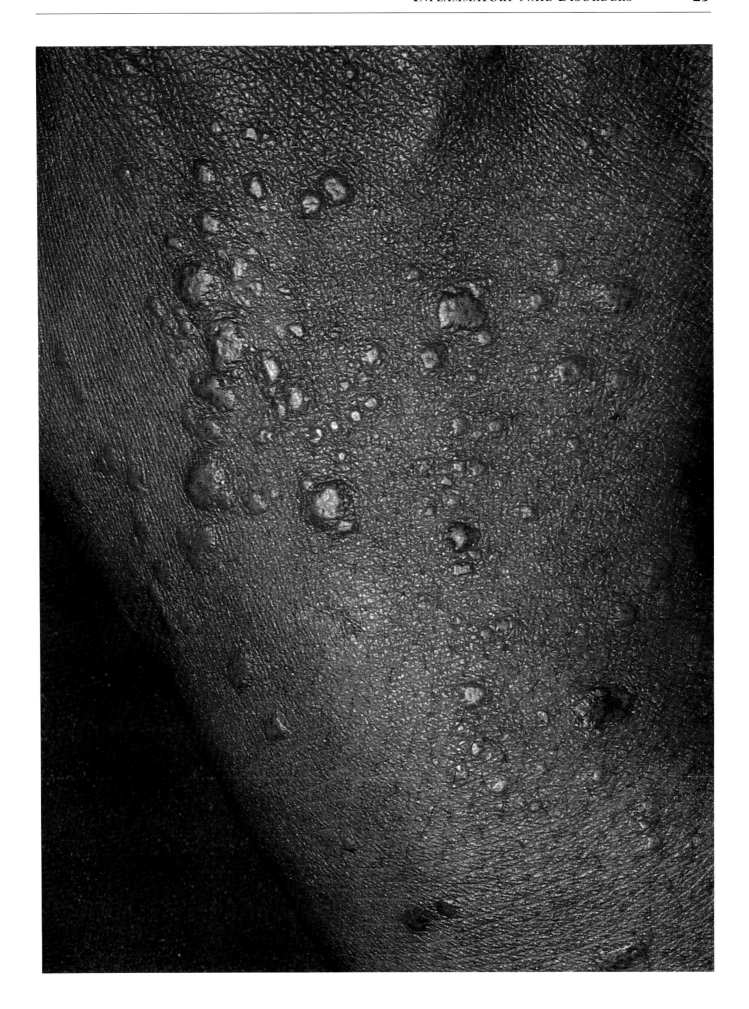

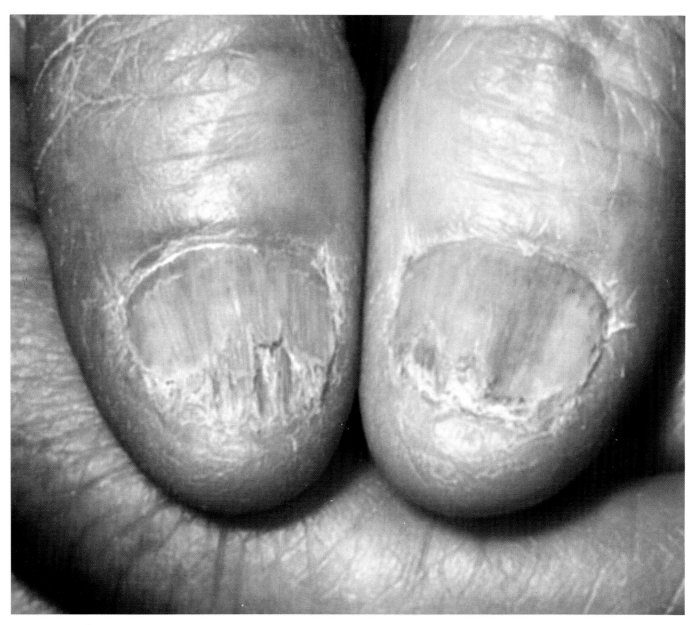

Figure 24. Onychorrhexis suggestive of lichen planus *(Courtesy of Boni E. Elewski, MD)*

INFLAMMATORY NAIL DISORDERS

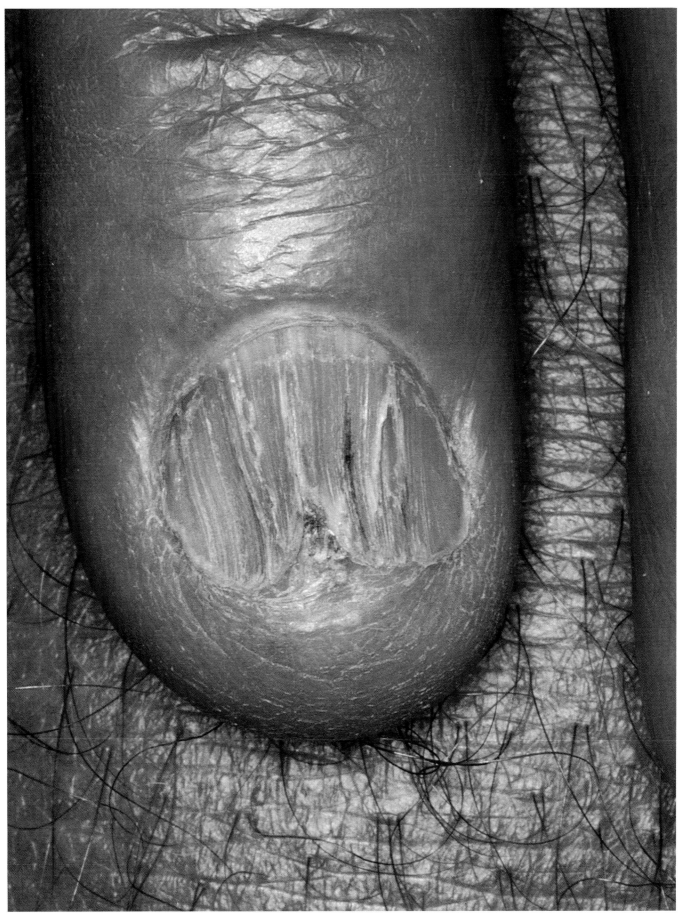

Figure 25. "Angel wing" deformity suggestive of lichen planus *(Courtesy of Gary Palmer, MD)*

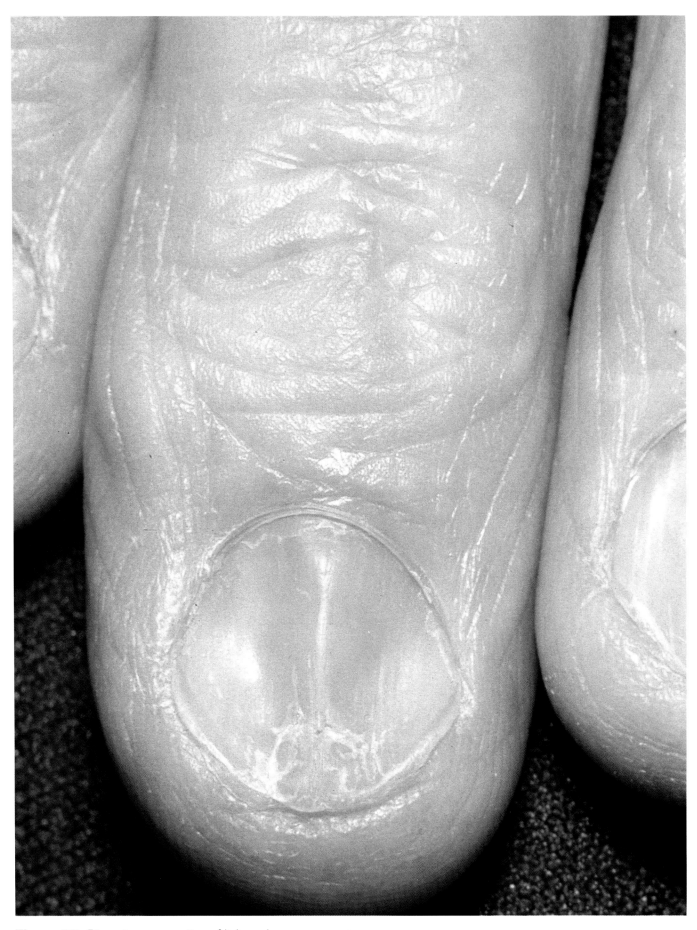

Figure 26. Pterygium suggestive of lichen planus *(Courtesy of Gary Palmer, MD)*

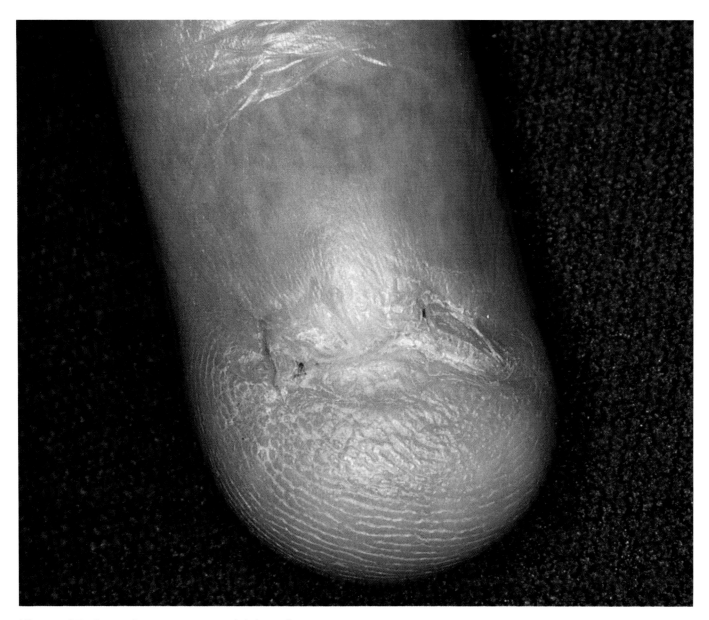

Figure 27. Anonychia in a patient with lichen planus *(Courtesy of Gary Palmer, MD)*

B.1.b. Differential Diagnosis

As was the case with psoriasis, lichen planus of the nails is relatively easy to diagnose when there is characteristic cutaneous or mucous membrane involvement. However, diagnosis is more difficult in the minority of cases where nail changes exist alone. A key clinical finding that differentiates lichen planus from onychomycosis is the presence of Wickham's striae, which may be in the mucous membranes (Figure 28). Pterygium is also not seen in onychomycosis or psoriasis. Patients with lichen planus may also develop the Koebner phenomenon, as is seen in psoriasis but not onychomycosis.

In addition to these clinical findings, mycologic studies should be considered if necessary to rule out onychomycosis using the same procedure described for diagnosing psoriasis of the nails (see pages 14-15.) It may also be necessary to perform an early nail biopsy to make a specific diagnosis, since extensive involvement of the nail apparatus may result in permanent damage. Table 4 lists some of the main histopathologic features of lichen planus in the nails; these are illustrated in Figures 29 through 32. Certain histopathologic findings are common to both the nails and the skin:[4] namely, hypergranulosis, hyperkeratosis, necrotic keratinocytes, irregular epidermal hyperplasia, coarse collagen bundles, dense lichenoid lymphohistiocytic infiltrates with melanophages, and vacuolar degeneration. However, those described in Table 4 are highly suggestive of lichen planus nail changes.

TABLE 4 — Common histopathologic characteristics of lichen planus in the nail [4]

- Diffuse hypergranulosis
- Marked compact orthokeratosis and focal parakeratosis
- Marked fibrosis in the papillary and reticular dermis
- Scarring may occur with resolution of condition
- Scale crust in cornified layer
- Vacuolar degeneration

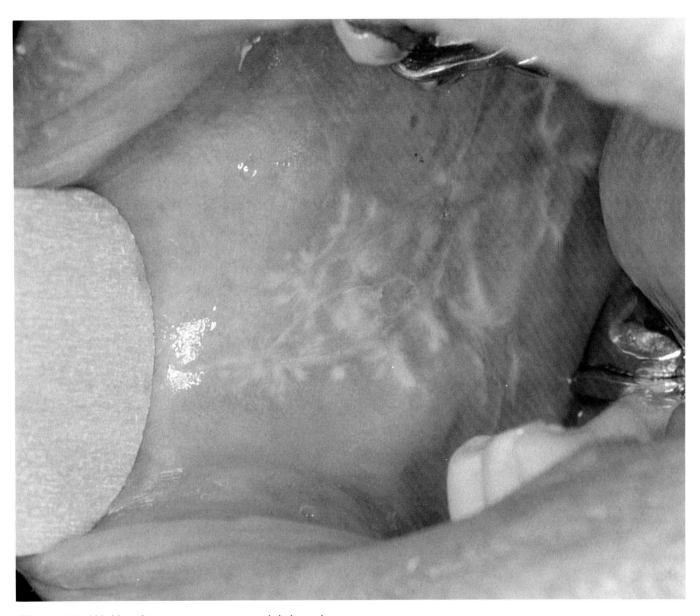

Figure 28. Wickham's striae in a patient with lichen planus *(Courtesy of Gary Palmer, MD)*

Figures 29–32. Histopathologic representations of lichen planus *(Courtesy of C.R. Daniel III, MD)*

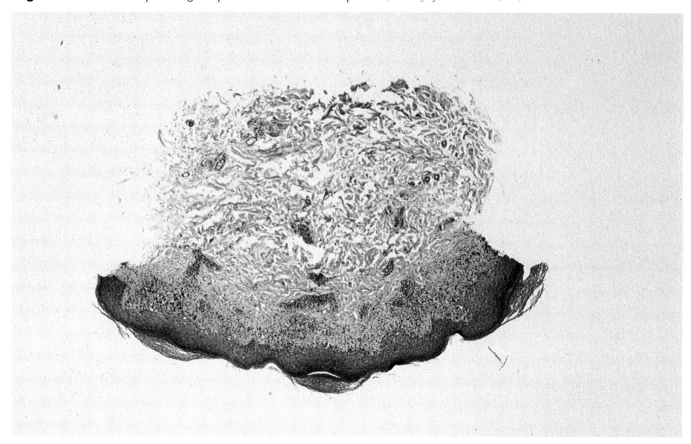

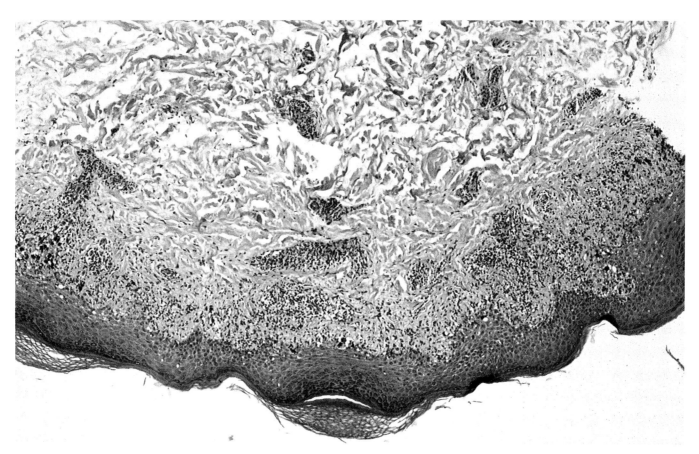

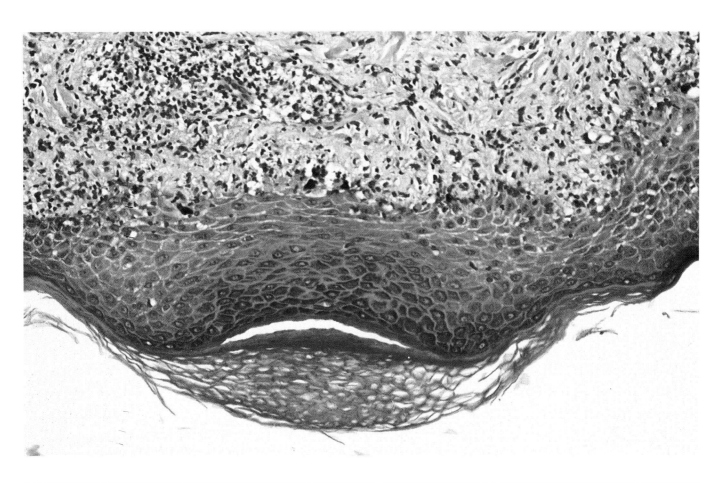

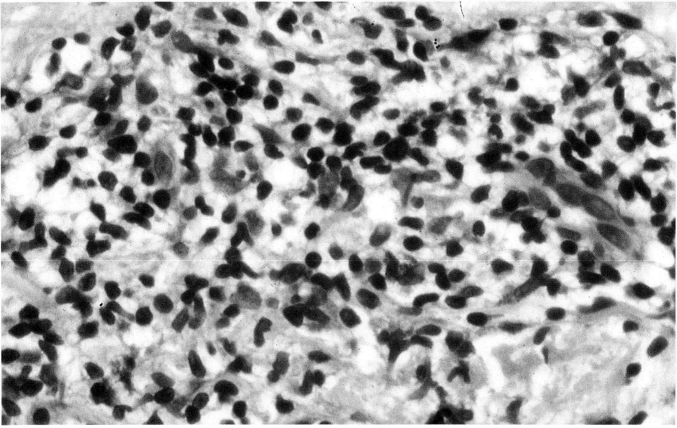

II.B.2. DARIER'S DISEASE

B.1.a. Clinical Presentation

Darier's disease, otherwise known as keratosis follicularis, is an inflammatory disorder that is inherited in autosomal dominant fashion with incomplete penetrance. The clinical manifestations of the disease reflect the anatomic site of involvement.[2] Nail involvement occurs in the vast majority of patients and is associated with characteristic distal wedge- or V-shaped subungual hyperkeratotic lesions, which are more likely to appear on the fingernails than the toenails (Figure 33).[2] Another characteristic finding in patients with Darier's disease is longitudinal streaks in the nails, which are initially reddish in color, but lighten over time (Figure 34).[7] Other possible nail changes include splinter hemorrhages, a spotty leukonychia, and thickening, splitting, or splintering of the nail plate.[2] In some cases, there is a spotted, reddish lesion in the lunula, which is distinct from the oil drop change. Secondary invasion by *Candida, Pseudomonas,* or dermatophytes is not uncommon.[2]

Other characteristics of Darier's disease include a thickened, and possibly malodorous, hyperkeratosis in the intertriginous folds and alopecia of the scalp, beard and other areas on the body. Flat keratotic papules generally appear in the nail folds and may also be detected at other cutaneous sites. Figures 35 and 36 demonstrate histopathologic representations of Darier's disease.

Figure 33. Distal wedge-shaped subungual hyperkeratosis suggestive of Darier's disease *(Courtesy of Gary Palmer, MD)*

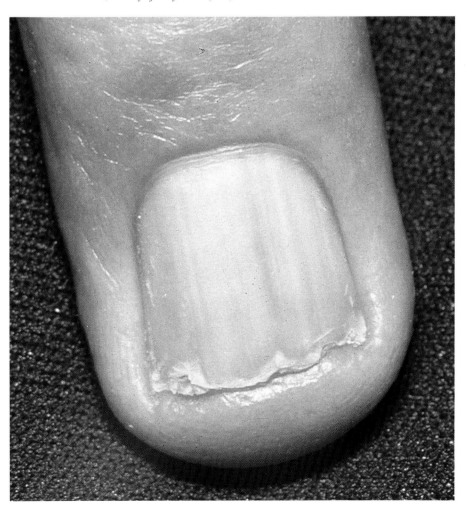

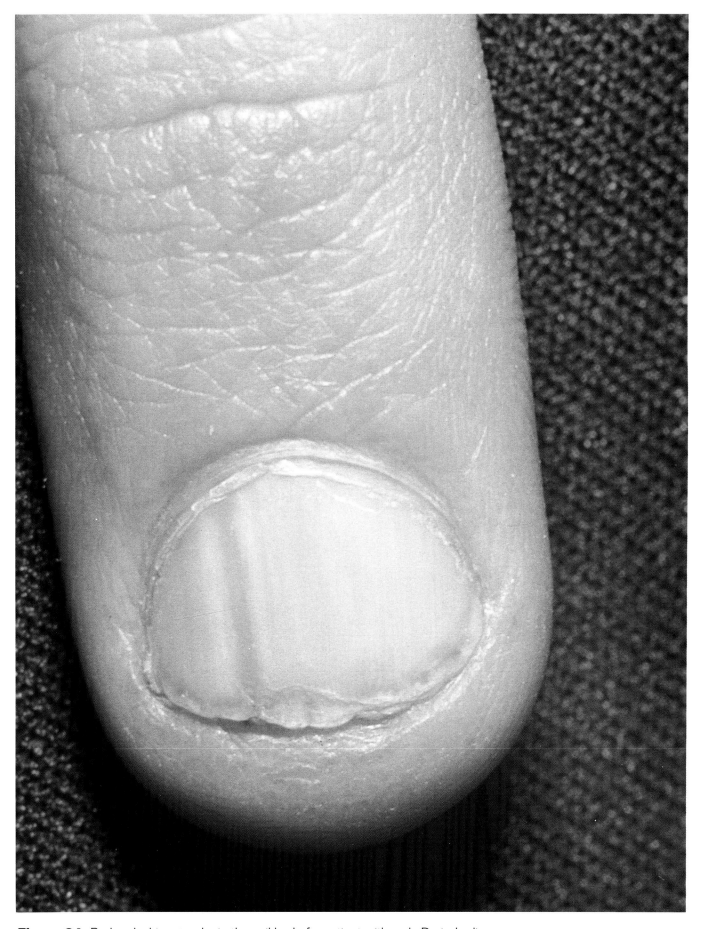

Figure 34. Red and white streaks in the nail bed of a patient with early Darier's disease *(Courtesy of Gary Palmer, MD)*

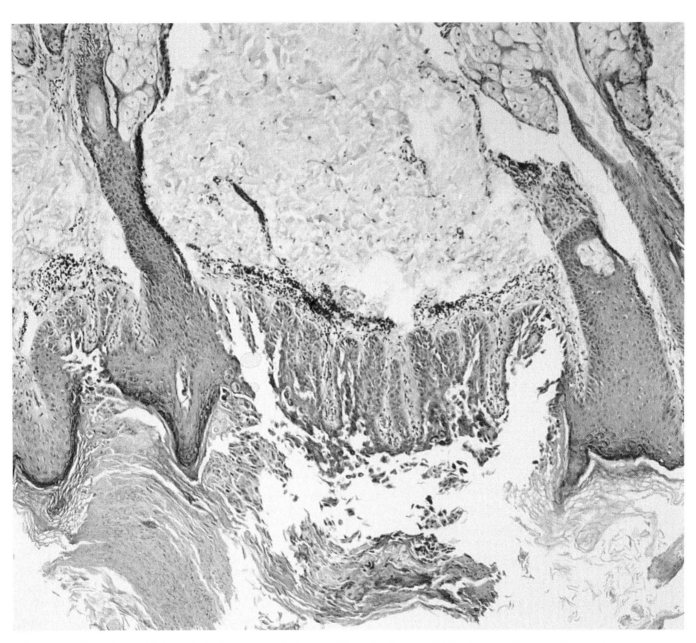

Figures 35 and 36. Histopathologic representations of Darier's disease (skin) *(Courtesy of C.R. Daniel III, MD)*

Inflammatory Nail Disorders 37

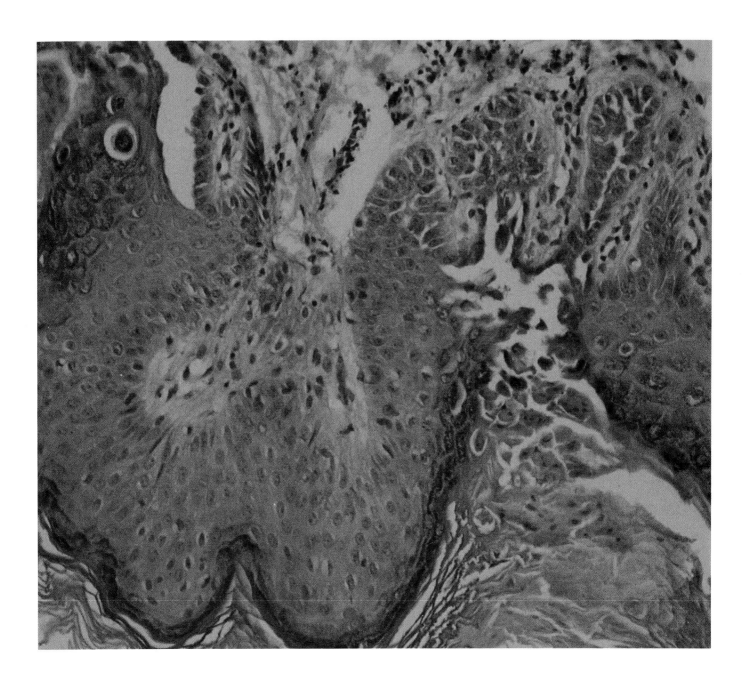

II.B.3. IDIOPATHIC TWENTY-NAIL DYSTROPHY

B.3.a. Description

Twenty-nail dystrophy is a nonspecific term used to refer to any disorder that affects all 20 nails. It has been observed in children and adults with alopecia areata, psoriasis, lichen planus, eczema, Darier's disease, and other dermatologic conditions, but may also occur in the absence of other systemic or cutaneous inflammatory disorders. Thus, it is unclear whether twenty-nail dystrophy is a distinct clinical entity or a clinical manifestation of lichen planus, psoriasis, or another disorder.

Figure 37. Fingernails in a patient with 20-nail dystrophy associated with lichen planus
(Courtesy of Gary Palmer, MD)

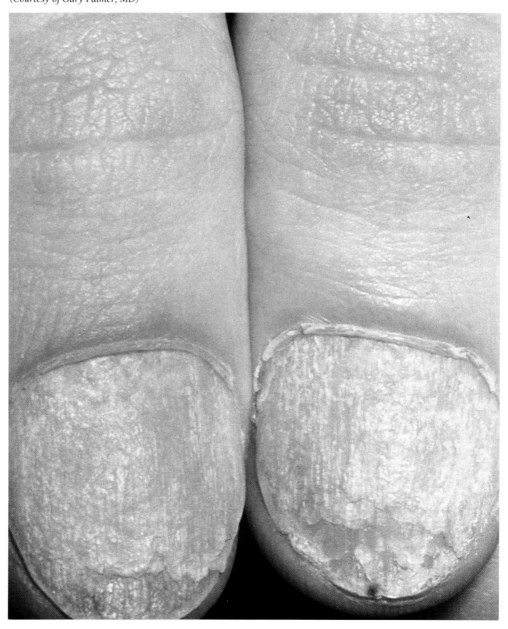

As previously discussed, fungal infections of the nail rarely affect both hands in healthy individuals. Therefore, involvement of all 20 nails usually rules out the possibility of onychomycosis in healthy individuals. Idiopathic 20-nail dystrophy of childhood is a condition that is usually self-limited. Figure 37 shows 20-nail dystrophy in the fingernails of a patient with lichen planus. As can be seen, the nail surface is dull and the nails are ridged and frayed at the edges. Figure 38 shows a toenail in a patient with idiopathic 20-nail dystrophy.

Figure 38. Toenail in a patient with idiopathic 20-nail dystrophy
(*Courtesy of Gary Palmer, MD*)

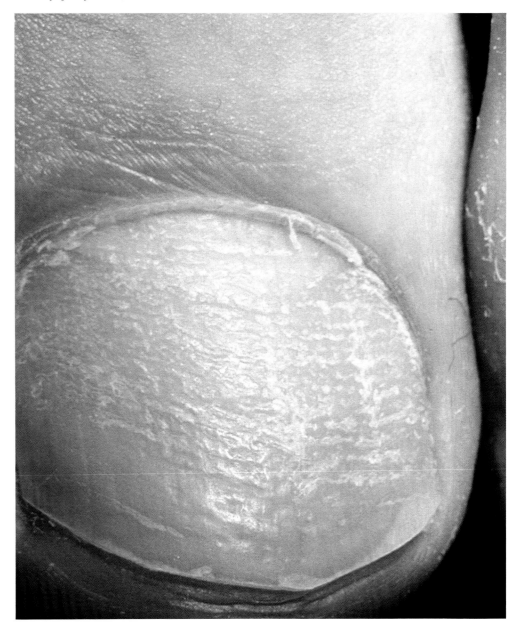

II.B.4. ECZEMATOUS DERMATITIS

B.4.a. Clinical Presentation

There are many clinical forms of eczema (e.g., atopic, nummular, dyshidrotic, contact), all of which can affect the nail. Patients with eczema often have history of allergies and asthma. They may also have sensitive skin or eczema elsewhere on their bodies. Those with contact dermatitis resulting from occupational or cosmetic causes are likely to have eczema on their hands or feet. However, fingernails are affected much more commonly than toenails, since the hands are exposed to the environment. Table 5 lists the myriad nail changes that may be associated with eczematous dermatitis. Figure 39 illustrates some of the nail changes characteristic of this condition.

TABLE 5 — Nail changes associated with eczematous dermatitis[2]

- Roughness, thickening, or brittleness of the nail plate
- Yellow discoloration
- Coarse, irregular pitting (rare)
- Transverse ridging
- Furrowing
- Smooth, shiny surface
- Nail shedding
- Onycholysis
- Subungual hemorrhages
- Chronic paronychia
- Transverse leukonychia
- Splinter hemorrhages

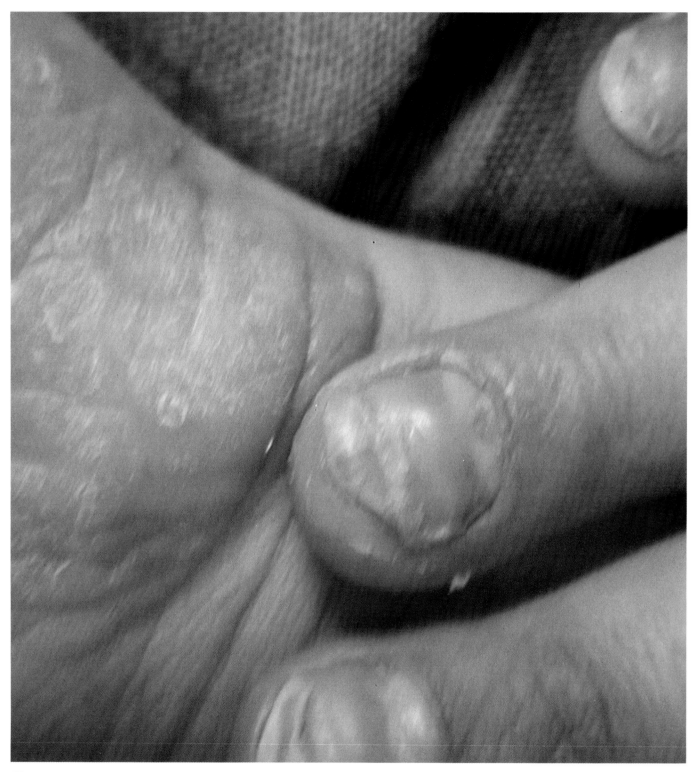

Figure 39. Nail changes suggestive of eczematous dermatitis *(Courtesy of C.R. Daniel III, MD)*

B.4.b. Differential Diagnosis

It is important to obtain a patient history to determine the cause of eczematous dermatitis and the course of onset. The main clue which suggests that the condition is contact dermatitis, and not onychomycosis, is the involvement of both hands. The presence of eczema in the periungual area or surrounding skin also can be used to help rule out onychomycosis.

II.B.5. SIMPLE ONYCHOLYSIS

B.5.a. Clinical Presentation

Onycholysis is a common condition seen in dermatologic practice and may be caused by numerous exogenous, endogenous and idiopathic, acquired or inherited factors (Tables 6 and 7). A prerequisite for the development of this condition is the disruption of the poorly-defined cement substance that binds the nail plate to the surrounding structures. Most commonly, onycholysis may be a clinical feature of a variety of dermatologic and systemic disorders.

Among the most common acquired causes of onycholysis are contact irritants and/or moisture, which affect the fingernails more frequently than the toenails. Those patients who are most likely to be affected are those who do extensive cleaning, bartenders, mechanics, construction workers, cosmetologists, and so forth. Figure 40 shows onycholysis secondary to frequent moisture.

TABLE 6 — Acquired causes of onycholysis[8]

Exogenous

- Contact/irritants/moisture
  - Nail cosmetics
  - Soaps/detergents
  - Raw foods (especially citrus)
  - Hydrocarbons
  - Cement

- Trauma

- Organisms
  - Dermatophytes
  - Other fungal organisms
  - Bacterial organisms
  - Viral organisms
  - Scabies

- Drugs and ingestants

Endogenous

- Dermatologic
  - Psoriasis
  - Lichen planus
  - Reiter's disease
  - Pemphigus
  - Hyperhidrosis

- Neoplasm
  - Benign
  - Malignant

- Systemic Diseases
  - Amyloid
  - Anemia
  - Peripheral ischemia
  - Lupus erythematosus
  - Scleroderma
  - Thyroid disease

INFLAMMATORY NAIL DISORDERS 43

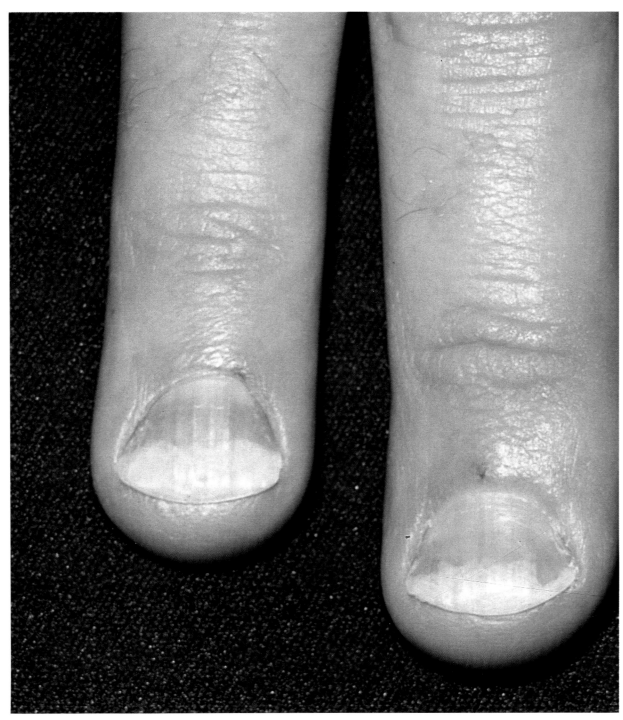

Figure 40. Onycholysis caused by exposure to moisture *(Courtesy of Gary Palmer, MD)*

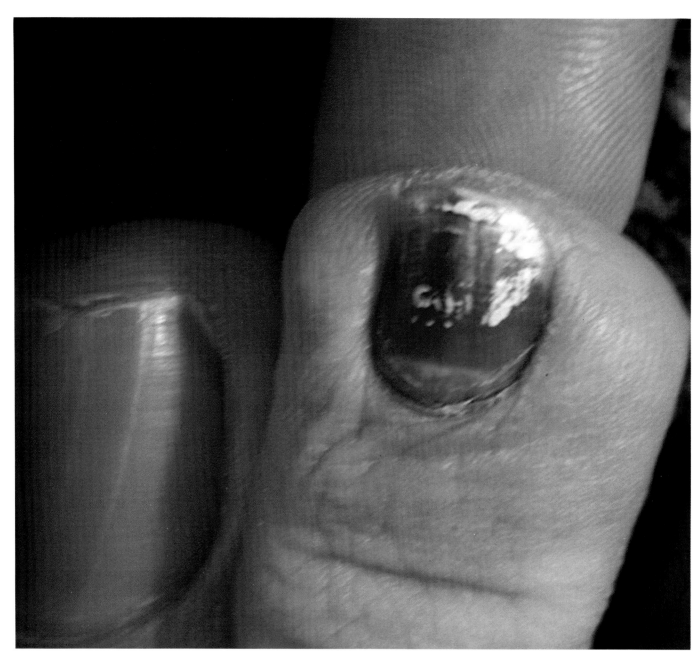

Figure 41. Onycholysis and hematoma of the toenail secondary to trauma *(Courtesy of C.R. Daniel III, MD)*

Trauma is another common cause of onycholysis, especially in toenails. Activities that involve rapid starts and stops, such as tennis, basketball, racquetball, long-distance running and other sports, may cause the toenails to hit the shoe, leading to a loosening of the connection between the nail plate and the nail bed (Figure 41). Other factors that may cause toenail onycholysis are endogenous conditions that alter the foot strike (e.g., edema, arthritis) or the wearing of high-heeled or pointed-toe shoes. There are also many traumatic causes of fingernail onycholysis, such as closing a finger in a door, hitting the fingernail with a hammer, or using the nail as a tool. As discussed later in this text (see "Distal Subungual Onychomycosis"), organisms such as dermatophytes are another common cause of onycholysis.

B.5.b. Differential Diagnosis

The diagnosis of simple onycholysis is predicated on obtaining a thorough patient history. The key factors that can be used to differentiate simple onycholysis from onycholysis caused by psoriasis are a lack of psoriasis elsewhere on the body, lack of pitting, oil drop change, and lack of subungual hyperkeratosis. Laboratory tests that can be used to distinguish between simple onycholysis and onychomycosis are described in the chapter on "Onychomycosis."

TABLE 7   **Inherited causes of onycholysis[8]**

Idiopathic
- Darier's disease
- Pachyonychia congenita
- Hidrotic ectodermal defect
- Hyperpigmentation, hypotrichosis, and dystrophy of the nails
- Hypoplasma enamel-onycholysis-hypohidrosis syndrome
- Leprechaunism
- Periodic shedding

CHAPTER THREE

TRAUMATIC NAIL DISORDERS

III.A. ONYCHOGRYPHOSIS

Onychogryphosis, otherwise known as "ram's horn nail," is an extreme example of unbridled nail growth.[9] It is a condition in which the nails become thickened and enlarged in response to trauma (e.g., wearing occlusive footwear), causing the nail to curve away from the source of pressure (Figure 42).[9] This condition typically develops in middle-aged or elderly persons, many of whom have bunions or wear pointed-toe shoes. In children, onychogryphosis is often a sign of neglect or may be associated with congenital malalignment, trauma, or external pressure from a first pair of poorly fitting shoes.[10] Onychogryphosis typically develops over the course of many years and may be associated with a dermatophyte infection.[9] Such nail growth may make it difficult for an individual to walk normally or to wear shoes.[9]

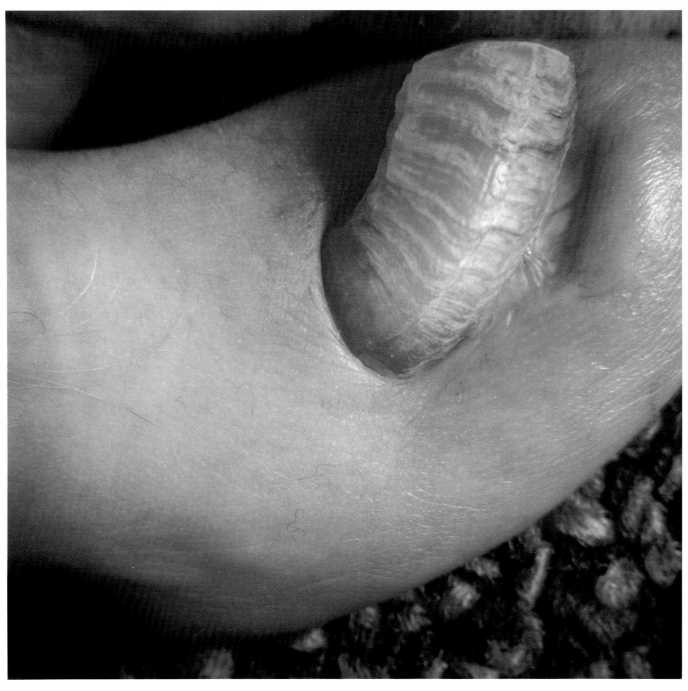

Figure 42. Onychogryphosis or "ram's horn nail" *(Courtesy of C.R. Daniel III, MD)*

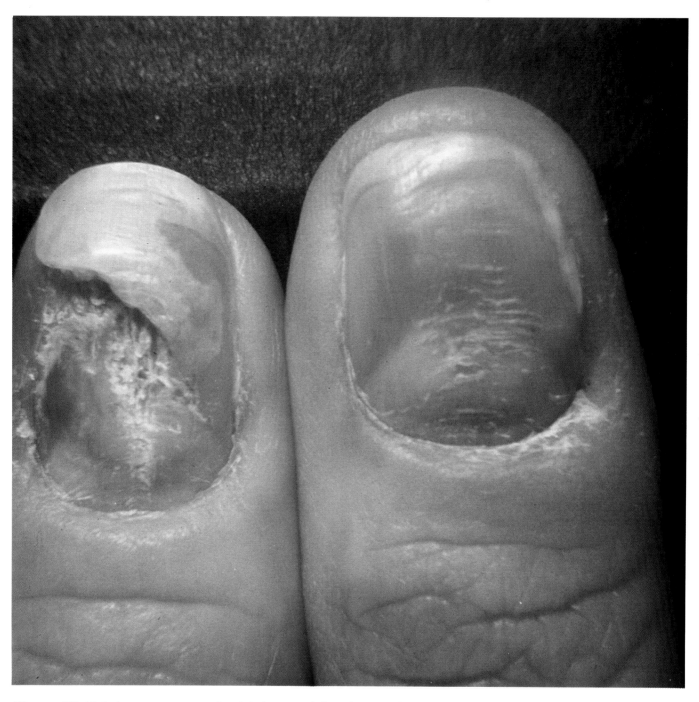

Figure 43. Nail changes associated with habit tic nail disorder (onychotillomania) *(Courtesy of C.R. Daniel III, MD)*

III.B. HABIT TIC NAIL DISORDERS

Habit tic nail disorders (e.g., onychotillomania, onychophagia) often manifest as a median furrow or depression in the middle of the nail, which develops when a patient picks at the nail cuticle (Figure 43). This condition is seen more commonly in fingernails than in toenails, and is usually limited to several nails on the nondominant hand. However, all 10 fingernails may be affected in patients with severe nervous disorders. The condition usually resolves if the patient is prevented from engaging in this activity. Both a KOH prep and fungus culture are negative.

III.C. TRANSVERSE STRIATE LEUKONYCHIA

Transverse striate leukonychia may be caused by either trauma or systemic factors. In both cases, a characteristic white line appears as the nail plate grows out.

Figure 44. Transverse striate leukonychia caused by trauma *(Courtesy of C.R. Daniel III, MD)*

III.C.1. Traumatic Transverse Striate Leukonychia

Traumatic transverse striate leukonychia may occur when a patient picks at the proximal nail fold or undergoes overly zealous manicuring. Picking at the nail around the cuticle can transmit a force down into the matrix, which disrupts normal keratinization and results in a white line that follows the contour of the proximal nail fold. However, the line is not homogeneous and does not transverse the entire width of the nail plate (Figure 44) (Daniel III, CR, personal communication).

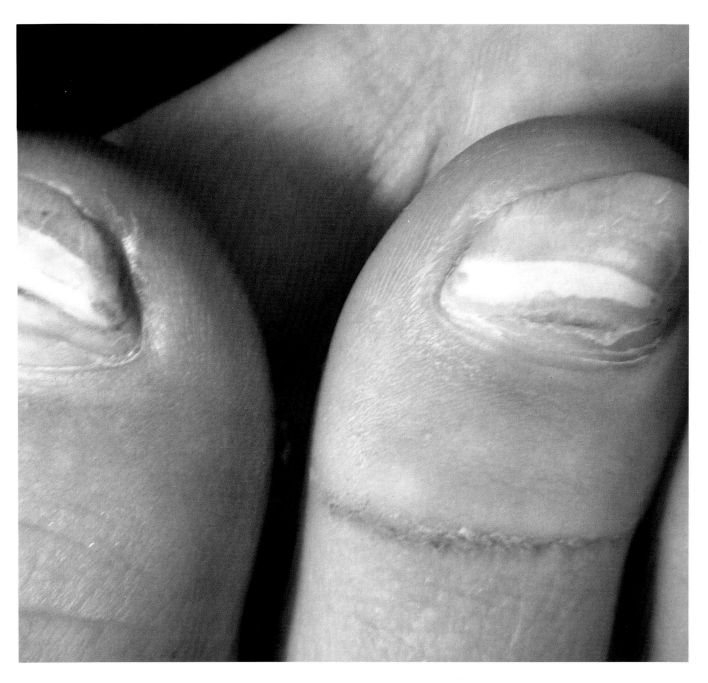

III.C.2. Nontraumatic Transverse Striate Leukonychia (Mee's lines)

Mee's lines are a form of nontraumatic transverse striate leukonychia that are classically associated with arsenic intoxication. As shown in Table 8, they may also result from virtually any other systemic insult. In contrast to the traumatic form of transverse striate leukonychia, Mee's lines are homogeneous, entirely white and traverse the entire nail plate, following the contour of the lunula (Figure 45).

TABLE 8	Some common causes of Mee's lines[11]
	• Arsenic poisoning[12–14]
	• Cardiac failure[12]
	• Childbirth[15]
	• Hodgkin's disease[12]
	• Myocardial infarction[12]
	• Pneumonia[12]
	• Psoriasis[12]
	• Renal failure[12, 16–18]
	• Sickle cell anemia[12]
	• Thallium poisoning[12]

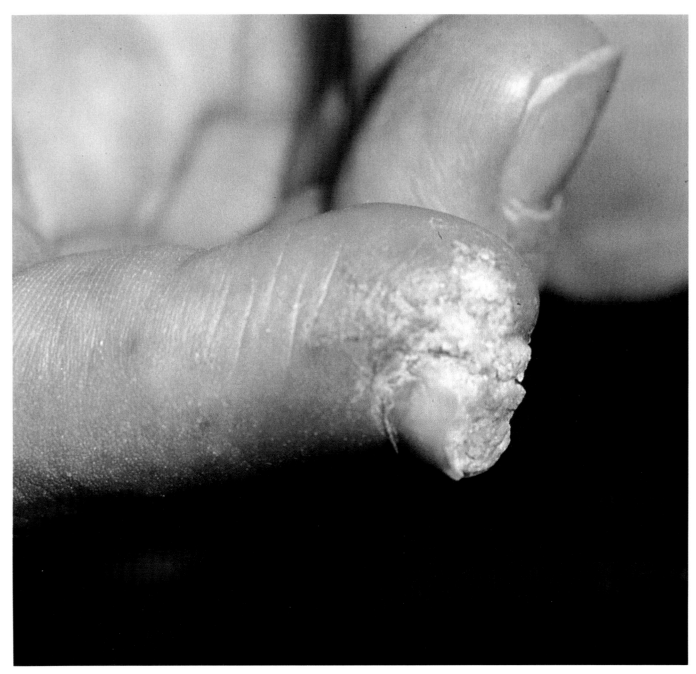

Figure 45. Transverse striate leukonychia caused by systemic factors *(Courtesy of C.R. Daniel III, MD)*

III.D. LEUKONYCHIA PUNCTATA

Leukonychia punctata (Figure 46) usually results from isolated trauma, such as tapping of the nails or overly aggressive manicuring. Contrary to earlier belief, it is not associated with a zinc deficiency.

Leukonychia punctata and striate leukonychia may resemble incipient proximal white subungual onychomycosis, which should be included in the differential diagnosis. Clues that this is the correct diagnosis are: history of trauma, negative KOH and fungal culture, and lack of fungus elsewhere (even though this could co-exist with fungus infections).

Figure 46. Leukonychia punctata *(Courtesy of C.R. Daniel III, MD)*

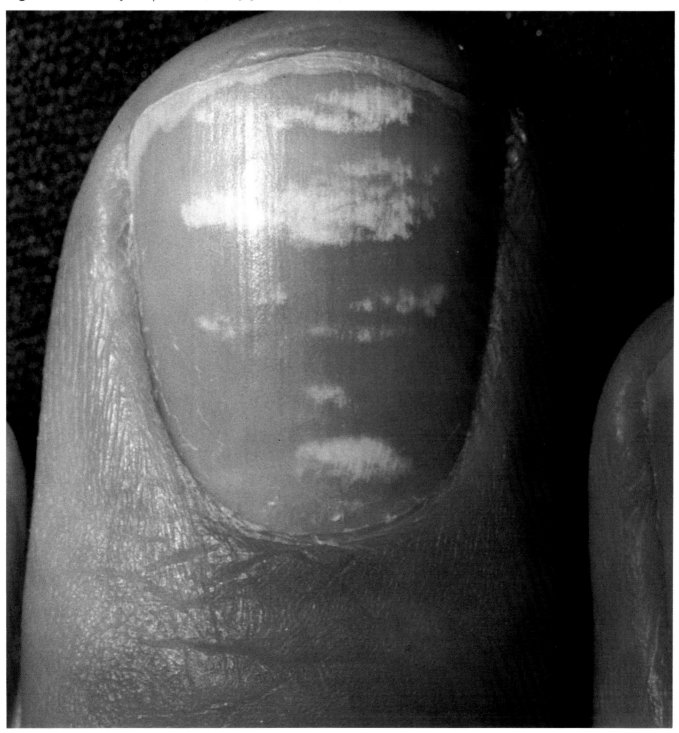

CHAPTER FOUR

LOCAL INFECTIONS

IV.A. ONYCHOMYCOSIS

Onychomycosis is one of the most common nail disorders seen by dermatologists.[19] Worldwide, the number of organisms that have been reported as capable of directly invading the nail and causing onychomycosis is increasing annually. This trend is likely to continue, given the growing number of elderly, the increasing use of potent antibacterial and chemotherapeutic agents, and the growing population of immunocompromised or immunosuppressed patients.

IV.A.1. Distal Subungual Onychomycosis

Distal subungual onychomycosis is the most common form of tinea of the nails (e.g., tinea unguium).[20] Approximately 90% to 95% of cases of distal subungual onychomycosis are caused by dermatophytes, primarily *Trichophyton rubrum* and *Trichophyton mentagrophytes*.[19]

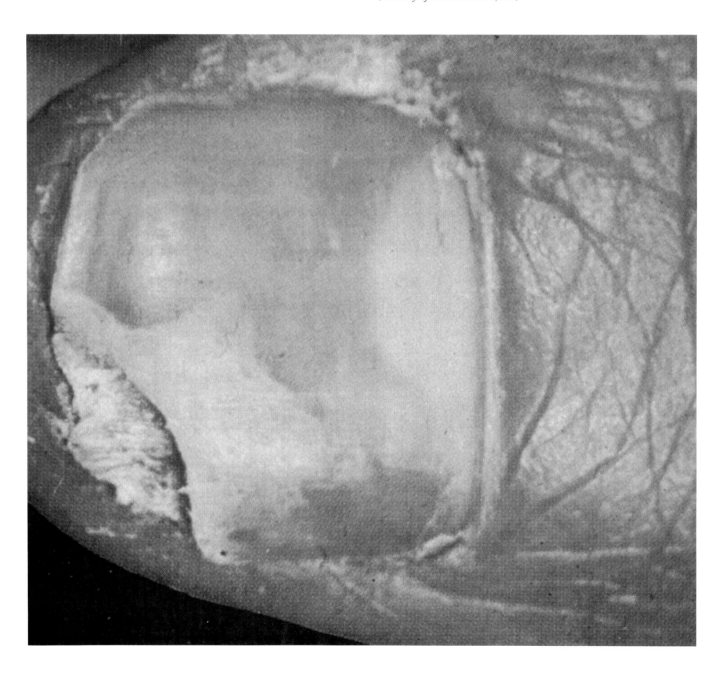

Figure 47. Distal subungual onychomycosis
(Courtesy of Boni E. Elewski, MD)

A.1.a. Clinical Presentation

The infection usually begins as tinea pedis, which is typically transmitted (physically) from parents to children, rather than being acquired in the locker room setting.

When the seal between the nail plate and the nail bed is broken by trauma or other causes, the organism penetrates the distal portion of the nail unit, around the lateral nail folds, and into the distal part of the hyponychial area. The early stages of the infection may be associated with yellowish or whitish discoloration of the nail, sometimes accompanied by splinter hemorrhages, onycholysis and isolated white or yellow islands in the nail plate. These manifestations become more pronounced as the infection progresses, leading to the characteristic changes shown in Figure 47.

A.1.b. Differential Diagnosis

To diagnose distal subungual onychomycosis, a curette, 2 mm dental spatula, or 2 mm nail elevator may be used to obtain crumbling subungual debris from under the distal edge of the nail. This material should then be submitted for studies. Diagnosis is predicated on a KOH stain and, ideally, fungal cultures using plain Sabouraud's dextrose agar containing an antibacterial agent (e.g., chloramphenicol) and Mycosel, or Sabouraud's agar containing both an antibacterial agent and cycloheximide.

Although some dermatophytes may be recognized by their characteristic colonies, a microscopic examination of an isolate is often required for confirmation. However, if it is not possible to obtain a fungal culture, a positive KOH stain for hyphal elements will adequately confirm a dermatophyte in more than 95% of cases. Multiple KOH stains and fungal cultures should be performed before onychomycosis is definitely ruled out.

The histologic diagnosis of onychomycosis is based on the clinical types of distal and proximal subungual onychomycosis, necessitating the presence of spores and hyphae in the superficial layer of the nail bed, and sometimes the deep portion of the nail plate, or the hyponychium. Biopsy specimens should be prepared with a PAS stain. Table 9 lists the histopathologic changes associated with subungual onychomycosis; the hyphae present in the compact horn are illustrated in Figure 48. In the absence of fungal elements, a diagnosis of psoriasis should be considered.

TABLE 9 Histopathologic features of subungual onychomycosis[6]

- Marked hyperkeratosis
- Hyphae present in compact horn
- Scale crust common
- Papillomatous epidermal hyperplasia common
- Extravasated erythrocytes usually in cornified layer

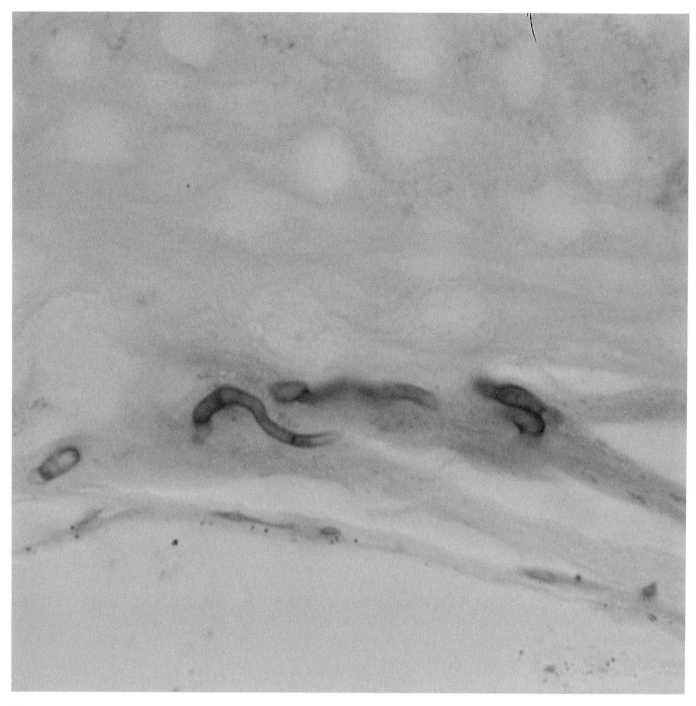

Figure 48. Hyphae present in compact horn *(Courtesy of C.R. Daniel III, MD)*

IV.A.2. Proximal Subungual Onychomycosis

Proximal subungual onychomycosis is a very uncommon form of onychomycosis, which is caused by dermatophytes (primarily *T. rubrum*) in about 90% to 95% of cases (Figure 49). The dermatophytes initially penetrate into the proximal nail fold region, thus creating the illusion that the infection is arising from beneath the nail cuticle. To obtain a specimen, a nail drill, or scalpel blade may be used to carefully obtain the debris. Identification of the organism is based on the same laboratory procedures described for diagnosing distal subungual onychomycosis.

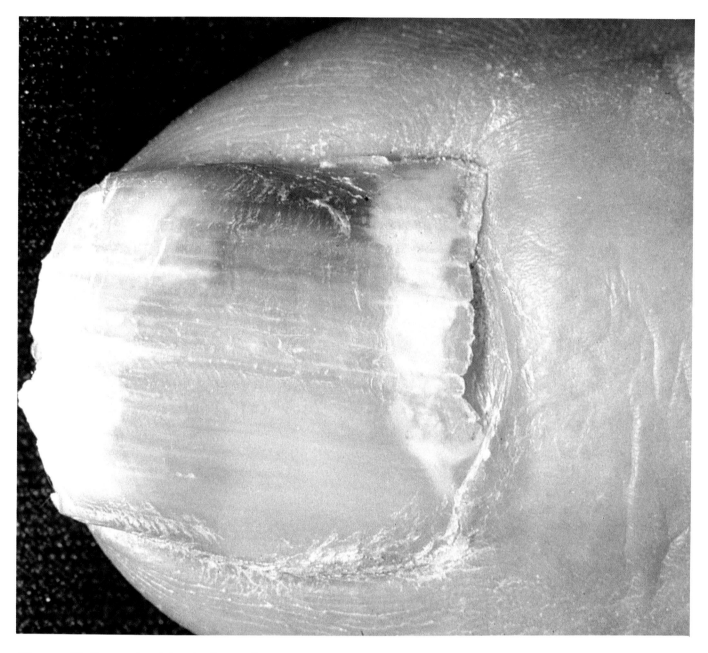

Figure 49. Proximal and distal subungual onychomycosis *(Courtesy of Gary Palmer, MD)*

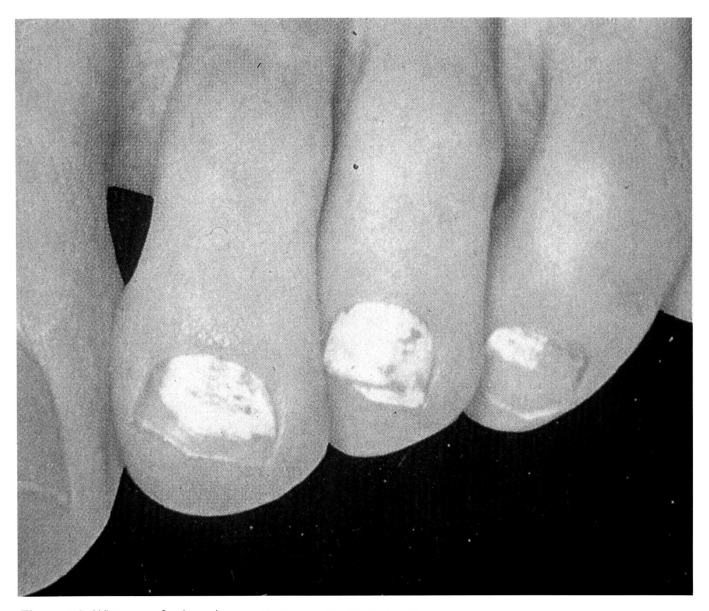

Figure 50. White superficial onychomycosis *(Courtesy of Boni E. Elewski, MD)*

IV. A.3. White Superficial Onychomycosis (Leukonychia Mycotica)

White superficial onychomycosis develops when the organism initially penetrates the nail plate (Figures 50 and 51). In contrast to distal and proximal subungual onychomycosis, white superficial onychomycosis is usually caused by *Trichophyton mentagrophytes*. This infection is common in toenails and is rare in fingernails.

Identification of the causative organism is based on the same procedure described for distal subungual onychomycosis. The histopathologic findings of white superficial onychomycosis are characterized by hyphae of dermatophytes or nondermatophyte fungi in the superficial portion of the nail plate, whereas the underlying nail bed is normal.

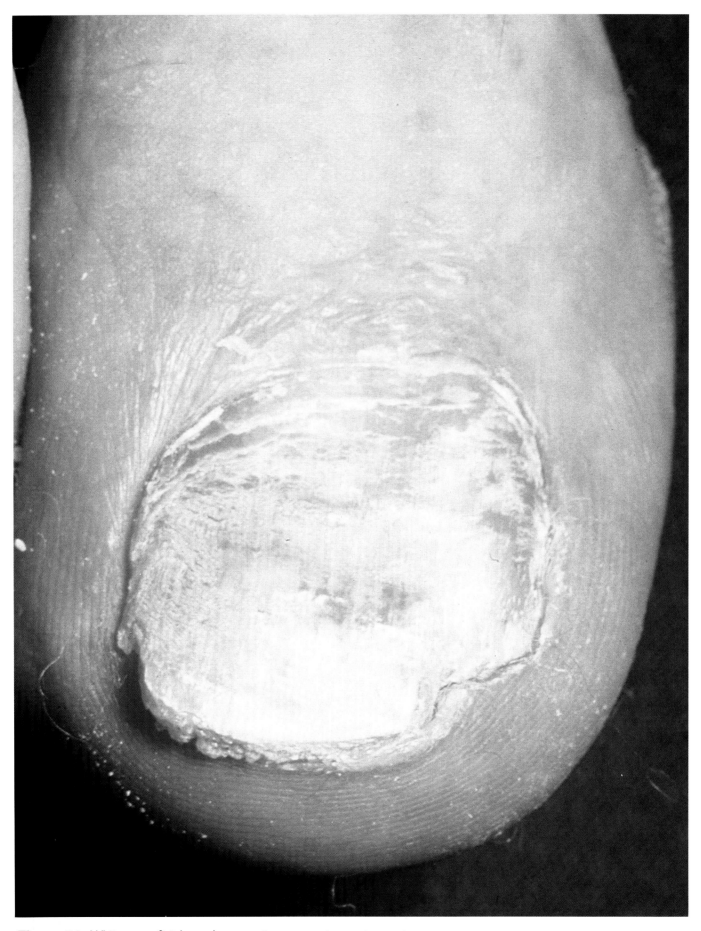

Figure 51. White superficial onychomycosis *(Courtesy of Gary Palmer, MD)*

IV.A.4. Total Dystrophic Onychomycosis

Total dystrophic onychomycosis can initially begin as distal subungual onychomycosis, proximal subungual onychomycosis, or superficial white onychomycosis, which then proceeds to involve the entire nail (Figure 52). Once the entire nail plate is affected, it is difficult to determine which form of onychomycosis caused the original nail infection.

IV.A.5. *Candida* Onychomycosis

In the United States, *Candida* sp. rarely cause primary infection in the immunocompetent host, but may exacerbate onycholysis. Secondary candidal nail infections in the immunocompetent host are only likely to develop when there is primary trauma to the nail, excessive exposure to moisture, or disruption of the nail fold. These conditions make it possible for *Candida* to inhabit the nail unit and to cause a secondary infection.

IV. B. PARONYCHIA

Paronychia is a common dermatologic condition characterized by redness, swelling and pain of the nail folds. It may be acute or chronic, infectious, or noninfectious. Exogenous factors play a greater role in its etiology than endogenous factors. Tables 10a. and 10b. list some of the diseases, conditions, and occupations associated with this disorder.

Figure 52. Total dystrophic onychomycosis *(Courtesy of Gary Palmer, MD)*

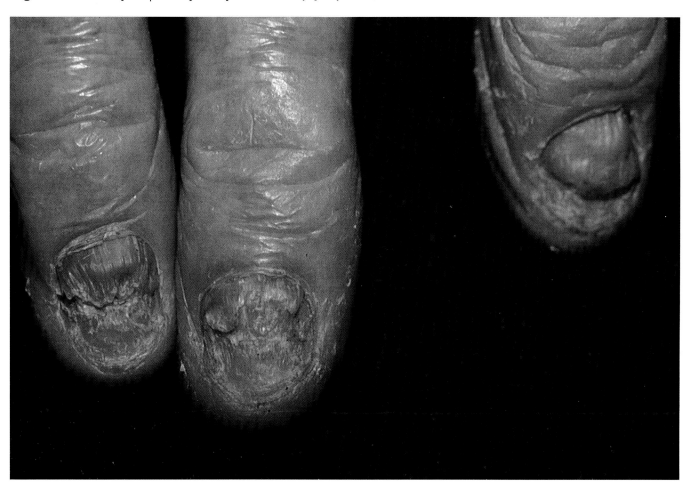

TABLE 10a.	Diseases/conditions associated with paronychia[21]

Bazex paraneoplastic syndrome[22]

Celiac disease[22]

Chronic lymphocytic leukemia[23]

Cold hands[20]

Diabetes mellitus

Dyskeratosis congenita[22]

Enchondroma[24]

Frostbite[25]

Histiocytosis X[26,27]

Hypoparathyroidism[22]

Ionizing radiation[28]

Leprosy[29]

Multicentric reticulohistiocytosis[31]

Parakeratosis pustulosa[28]

Pemphigus[31]

Pernio[25,32]

Progressive systemic sclerosis[33]

Psoriasis[34,35]

Raynaud's disease[25]

Reiter's syndrome[36]

Retinoids

Rubinstein-Taybi syndrome[37]

Sarcoid[22]

Stevens-Johnson syndrome[38]

Systemic lupus erythematosus[39]

Traumatic injury[25]

Tunga penetrans[28]

Varicose veins[32]

Vascular thrombosis[25]/thrombophlebitis[32]

Vasculitis[25]

Verrucous nevus/nevus unius lateris[40]

Zinc deficiency [41]

TABLE 10b.	Occupations associated with paronychia[21]

Barbers/hairdressers[42]

Bartenders[42]

Carpenters/builders

Cooks

Cosmetologists

Dentists/dental hygienists

Engravers/etchers/glaziers[28]

Fishermen

Gardeners/florists

Groundskeepers

Housewives

Janitorial/domestic workers[42]

Legume shellers[28]

Meat/raw food handlers

Mechanics

Oil-rig workers[43]

Painters[28]

Photographic/x-ray developers

Pianists[28]

Radio workers[28]

Secretaries (clerical)

Shoemakers[28]

Waitresses/waiters[23]

IV.B.1. Acute Paronychia

Acute paronychia is a bacterial infection that usually appears first on the dominant hand, particularly the thumb and index finger (Figure 53). It rarely affects the toenails. The most common cause is trauma to the nail unit, particularly the nail folds, resulting in disruption of the nail fold, the cuticle, and the nail plate. This, in turn, produces an environment conducive to colonization by bacteria such as *Staphylococcus aureus, Pseudomonas aeruginosa*, and A or D *streptococci*.[44] If left untreated, it may lead to dystrophy of the nail plate.

To diagnose acute paronychia, it is important to inquire about the nature of the patient's occupation, household activities and hobbies. A thorough history should be obtained, including information about the patient's health, particularly diabetes mellitus. The patient should also be questioned about psoriasis and any other forms of dermatitis. It may be valuable to review the list of factors associated with paronychia (see Tables 10a and 10b) with the patient and to identify possible predisposing causes of the infection.

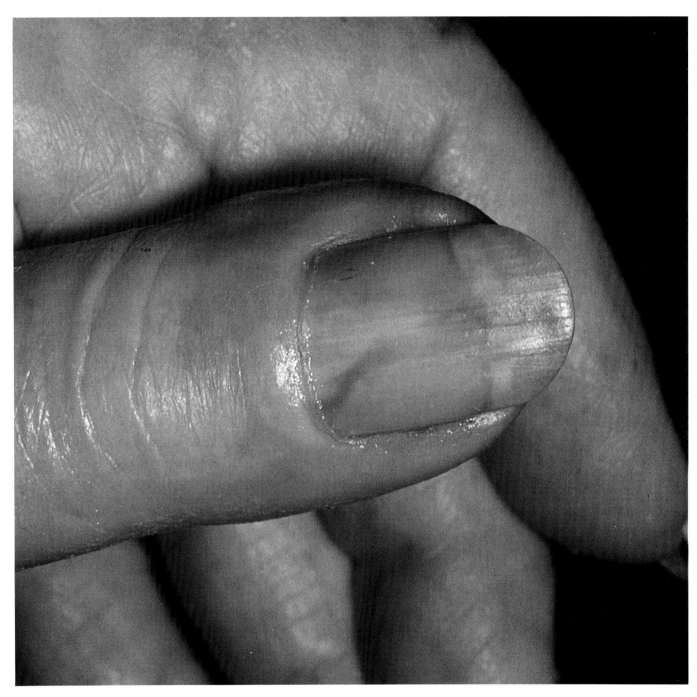

Figure 53. Acute paronychia *(Courtesy of Gary Palmer, MD)*

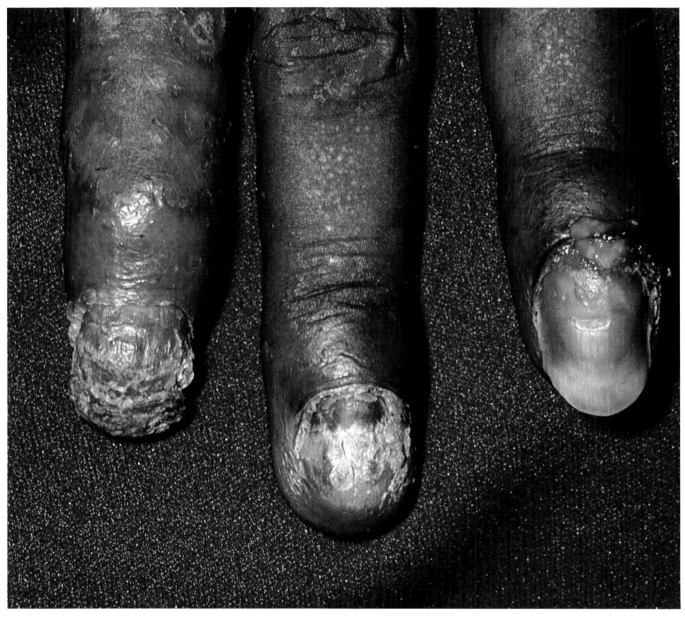

Figure 54. Chronic paronychia *(Courtesy of Gary Palmer, MD)*

IV.B.2. Chronic Paronychia

Chronic paronychia is usually a multifactorial condition in adults, which may be perpetuated by irritants, contactants, Candida and low-grade bacterial infection (Figure 54). It is associated with a reddened cuticle, brownish discoloration of the adjacent nail, chronic inflammation of the nail fold, and thickening of the nail plate. In extreme cases, there may be dystrophy of the nail plate, leading to the virtual disappearance of the nail cuticle.

IV.C. OTHER INFECTIOUS CONDITIONS AFFECTING THE NAILS

IV.C.1. Chronic Mucocutaneous Candidiasis

Chronic mucocutaneous candidiasis is a primary *Candida* infection in the immunocompromised host. While the main symptoms usually involve the mucocutaneous areas, there may be massive thickening of several or all 20 of the nails as well, due to invasion by *Candida albicans* (Figure 55).[10] Other nail changes may include paronychia, subungual hyperkeratosis, and onycholysis, which may progress to involvement of the entire nail plate and result in yellow-brownish discoloration, brittleness, chipping, and onychauxis.[10] Table 11 lists some of the conditions associated with chronic mucocutaneous candidiasis.

A KOH prep of nail scrapings from these patients reveals the characteristic *Candida* pseudohyphae.

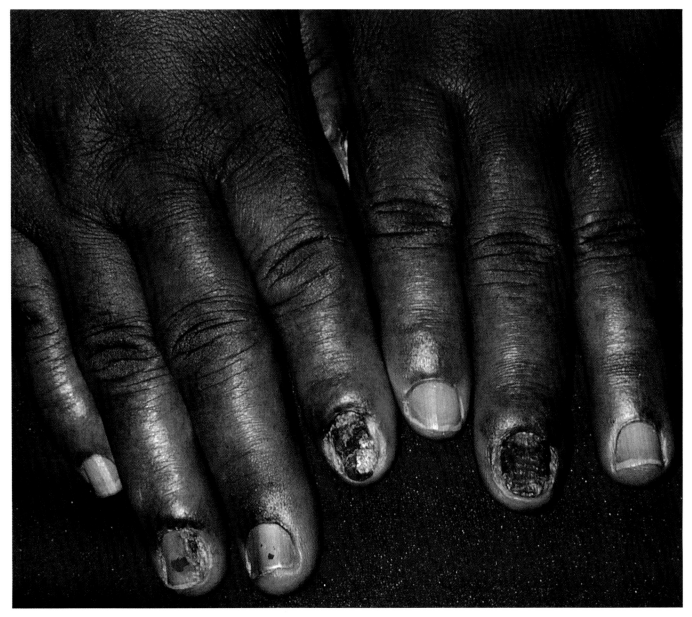

Figure 55. Nail changes in a patient with chronic mucocutaneous candidiasis *(Courtesy of Gary Palmer, MD)*

TABLE 11 — Conditions associated with chronic mucocutaneous candidiasis[10]

- Primary immunodeficiency
  - DiGeorge's
  - Nezelof's
  - Common variable immunodeficiency
  - Hyper IgE syndromes
  - Myeloperoxidase deficiency

- Secondary immunosuppression

- Medical conditions
  - Multiple endocrinopathies (e.g., parathyroid, adrenal, thyroid, diabetes)
  - Iron-deficient states
  - Thymoma, myasthenia, hypogammaglobulinemia
  - Ectrodactyly-ectodermal dysplasia cleft (EEC) syndrome
  - Dental enamel dysplasia syndrome

IV.C.2. Nail Changes in Patients with HIV Infection

The presence of onychomycosis has been helpful in classifying patients with HIV disease, since it generally correlates with the patient's immune status as reflected by his or her T-cell count.[45,46] Although the causative organisms and clinical presentations of onychomycosis in patients with HIV infection are similar to those in patients without HIV disease, there are certain important differences between these two populations.

First, HIV-positive patients commonly develop proximal white subungual onychomycosis in both the fingernails and toenails, while this is rarely the case in the general population, especially when there is fingernail involvement.[47] Thus, the presence of proximal white subungual onychomycosis of the fingernails or toenails should strongly suggest possible HIV disease.[47] Another key factor that can be used to identify onychomycosis in the immunosuppressed vs. the immunocompetent host is the infectious pathogen involved. When a healthy individual develops superficial white onychomycosis of the toenails, it is usually caused by *T. mentagrophytes*.[47] In contrast, most cases of proximal white subungual onychomycosis or superficial white onychomycosis in HIV-positive patients are caused by *T. rubrum*.[47]

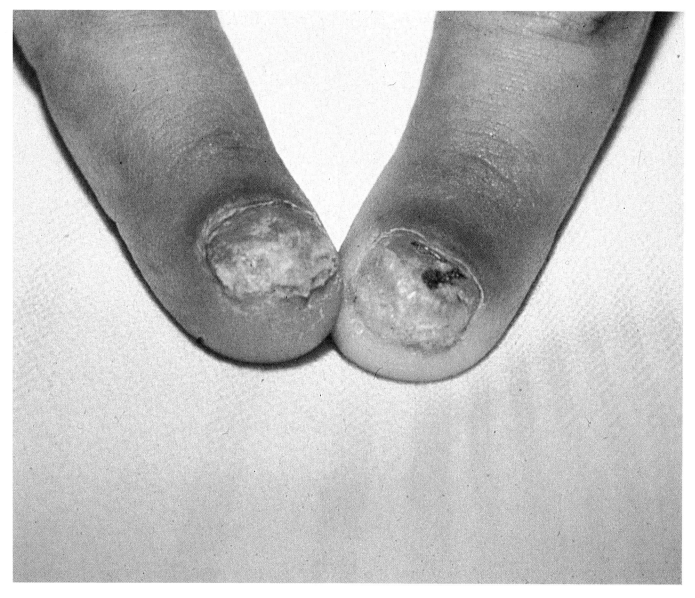

Figure 56. Hypertrophy of the nail bed caused by *Candida* infection in a HIV-positive patient *(Courtesy of Boni E. Elewski, MD)*

In addition, mycotic keratoderma is more common in HIV-infected patients than in the general population with "one hand, two feet tinea," and may affect both hands. Unlike classical onychomycosis, onychomycosis in the HIV-positive host may spread rapidly to all fingernails and toenails. It may also extend to the periungual region.[48]

Especially in cases of multiple nail involvement, *Candida* is the primary pathogen of the nail bed and plate in HIV-infected patients and may lead to *Candida* nail dystrophy in patients whose helper T-cell count is <100. Onycholysis or paronychia may occur, as well as a hypertrophic nail bed *Candida* infection (Figure 56).[49]

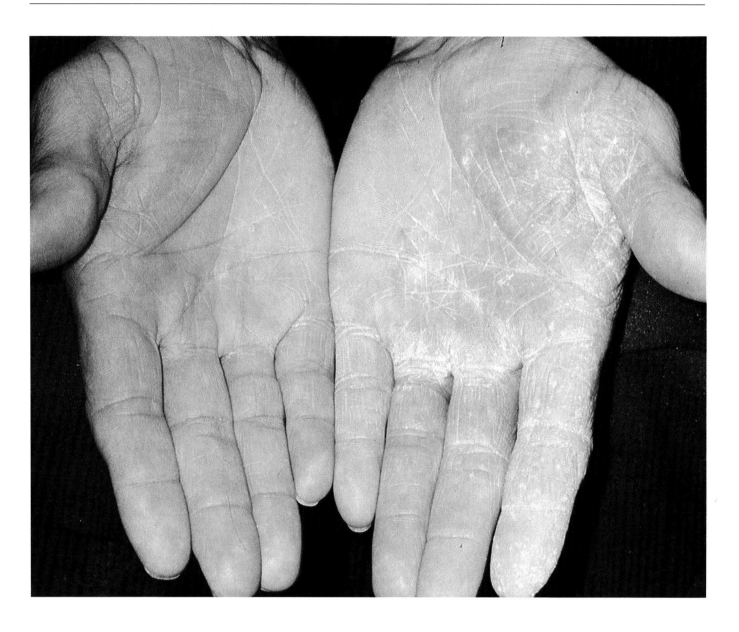

Figures 57 and 58. Mycotic keratoderma
(*Courtesy of Gary Palmer, MD*)

IV.C.3. Mycotic Keratoderma

Mycotic keratoderma is a fungal infection of the palm or sole (Figures 57 and 58). Although it is not unusual in the general population, it is more commonly seen in conjunction with proximal white subungual onychomycosis in HIV-positive individuals. In contrast to the findings in immunocompetent individuals, dermatophytes can be found in the periungual region in people with HIV disease. Therefore, HIV infection should be suspected if a patient presents with mycotic keratoderma on both hands, as well as proximal white subungual onychomycosis. The nails may act as a reservoir for the fungus to spread elsewhere.

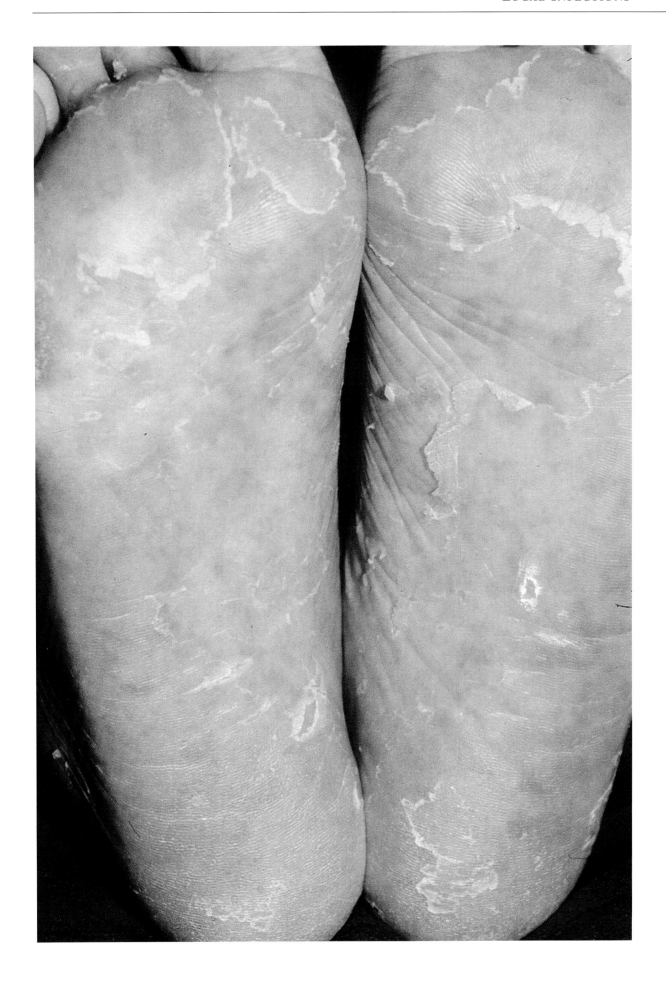

IV.C.4. Green Nail

There are two primary causes of the syndrome known as "green nail" (Figure 59): *Pseudomonas* sp. and, less commonly, *Candida* sp. When caused by pseudomonas, this condition is associated with greenish discoloration of an onycholytic nail, which occurs when the nail bed is stained with a pigment called pyocyanin. Although in many cases there is no active infection, bacterial and fungal cultures may be performed even when the pyocyanin pigment is still visible. Little is known about the pigment causing the green color occasionally seen with *Candida*.

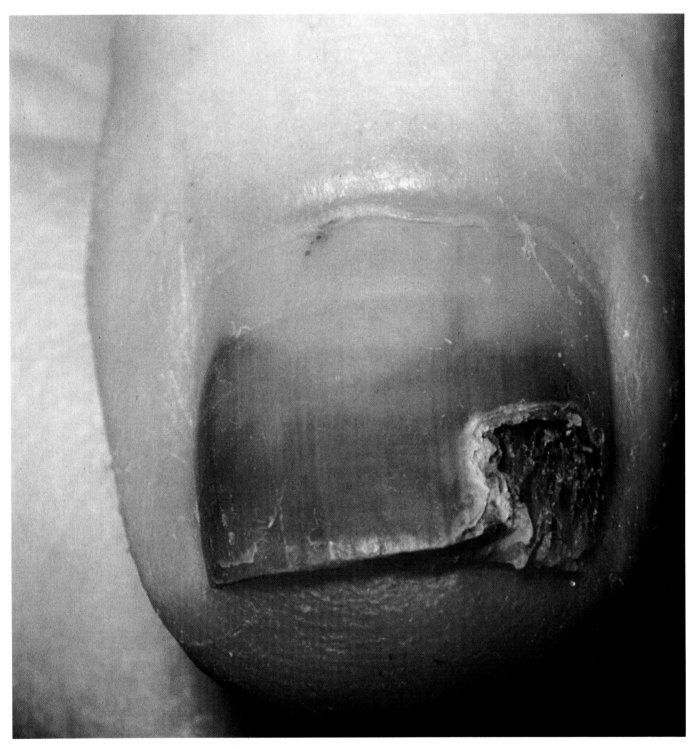

Figure 59. "Green" nail *(Courtesy of Gary Palmer, MD)*

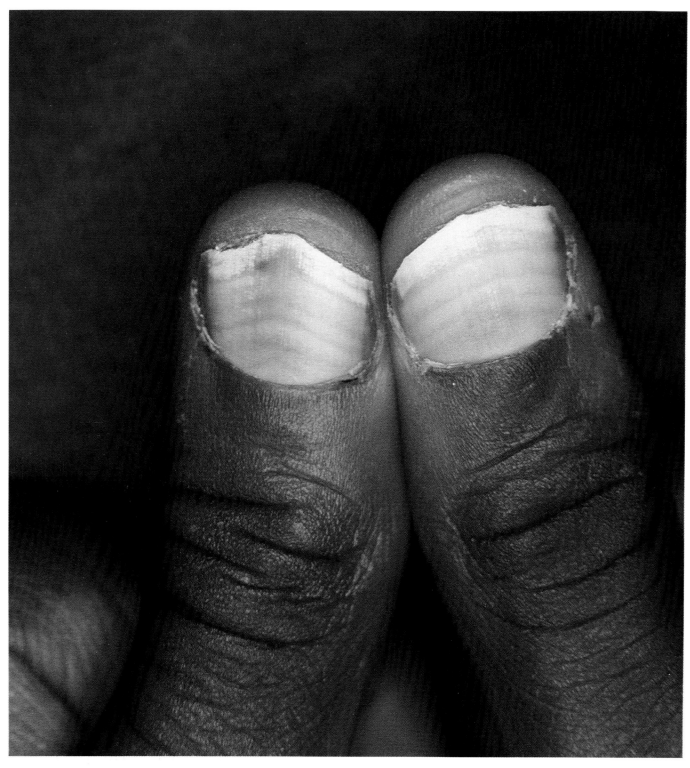

Figure 60. Crusted scabies *(Courtesy of Rosemary T. DePaoli, MD)*

IV.C.5. Crusted Scabies

Scabies are caused by the mite, *Sarcoptes scabiei* var. *hominis*, and commonly affect the finger web spaces, wrists, elbows, axillae, belt line, buttocks, nipple area (women), and genitals (men), but rarely the nails.[2] Crusted scabies, otherwise known as "Norwegian scabies," occur primarily in patients with Downs' syndrome, although they have also been reported in patients with HIV disease. They differ from most other forms of scabies, which do not involve the nail unit. This condition may be associated with hypertrophy of the nail plate and the presence of thousands of organisms in the nail plate, causing thickening of the nail.

Microscopic examination of biopsy specimens of the nail bed shows marked hyperkeratosis with focal parakeratosis. Ascarides can sometimes be seen in the cornified layer of the nail bed. The condition is highly contagious and is often associated with much itching. Crusted scabies (Figure 60) may be differentiated from onychomycosis by the presence of the former on the fingernails of both hands as well as pruritic papules scattered around the body.

GLOSSARY

Acuminate papule: Tapered collection of pus or scale within or beneath the skin, often in a hair follicle or sweat pore.

Alopecia areata: A microscopically inflammatory, usually reversible, patchy loss of hair, commonly occurring in sharply defined areas and usually involving the scalp or beard.

"Angel wing" deformity: A characteristic nail deformity associated with lichen planus in which the central portion of the nail appears raised and the lateral portions are depressed.

Anonychia: Absence of nail plate or nail unit.

Beau's lines: Transverse ridges that traverse the nail plate. They are caused by intermittent injury in the vicinity of the proximal nail fold, with resultant damage to the underlying proximal nail matrix. They are one of the most common, but least specific, nail changes associated with systemic disease.

Extravasated erythrocytes: Red blood cells that are found outside of blood vessels.

Green nail: Infection characterized by greenish color and caused by bacteria and/or fungus.

Hypergranulosis: Increased thickness of the granular layer of the epidermis, associated with hyperkeratosis.

Hyperkeratosis: Hypertrophy of the superficial layer of the skin.

Hyphae: Filaments or threads that compose the mycelium of a fungus.

Hyponychium: The thickened epidermis underneath the free distal end of the nail.

Hypotrichosis: Presence of less than the normal amount of hair.

Keratinocyte: The epidermal cell which synthesizes keratin. Constitutes 95% of the epidermal cells and, together with melanocytes, forms the binary cell system of the epidermis.

Koebner phenomenon: When the skin or nails are traumatized (as in psoriasis or lichen planus), a psoriatic or lichen planus lesion occurs in that spot several weeks later.

Leukonychia: Real or apparent whitening of the nail plate.

Leukonychia punctata: Small round, whitish discolorations in the nail plate that grow out with the nail plate.

Lichen planus: An inflammatory, pruritic cutaneous disease, sometimes also involving the oral and genital mucosa scalp and nails.

Lichenoid lymphohistiocytic infiltrates: Lymphocytes and histocytes invade the dermo-epidermal junction.

Lunula: The half-moon, white area at the base of the nail.

Mee's lines: Transverse white bands that tend to occur on several nails at once and usually spread across the entire breadth of the nail plate. They tend to be relatively homogeneous and to have smooth borders; in addition, they tend to follow the contour of the distal lunula, with a rounded distal edge.

Melanophage: A macrophage laden with phagocytosed melanin.

Mucocutaneous: Pertaining to, or affecting, the mucous membrane and the skin.

Nail bed: The structure which supports the entire nail plate extending from the nail matrix to the hyponychium.

Nail folds: Consist of one proximal and two lateral; define and provide support for the nail unit, as well as direct the growth process of the nail plate.

Nail matrix: The thick portion of the nail bed beneath the proximal nail fold from which the nail develops.

Nail plate: Horny, cutaneous portion of the dorsal surface of the distal end of a finger or toe.

Nail unit: Entire nail structure which is comprised of nail bed, nail folds, nail matrix, and nail plate.

Oil drop sign: Lesion in the nail bed, which is salmon-colored and resembles a spot of oil. Caused by the exudation of a serum glycoprotein when psoriasis is present.

Onychauxis: Simple hypertrophy of the nail plate without deformity.

Onychogryphosis: Hypertrophy of the nail plate which produces a hornlike deformity known as "Ram's horn nail." The condition is most likely due to trauma. One side of the nail appears to grow faster than the other, resulting in a curvature of the nail plate away from the site of trauma.

Onycholysis: Distal separation of the nail plate from the underlying and/or lateral supporting structures (e.g., the nail bed).

Onychomycosis: Fungal infection of the nail unit; tinea unguium.

Onychophagia: Nail biting.

Onychorrhexis: Longitudinal striations of the nail plate, usually superficial, which result in abnormal brittleness of the nails.

Onychotillomania: Compulsive habit of tearing, picking, or destroying nails or cuticle by some method.

Oral leukokeratosis: Whitish change in the mucus membrane.

Orthokeratosis: Hyperkeratosis without parakeratosis.

Pachyonychia: Partial or complete thickening of the nail plate.

Parakeratosis: Persistence of the nuclei of the keratinocytes into the horny layer of the skin.

Parakeratosis pustulosa: A benign, inflammatory condition of the distal phalanx that occurs in children.

Paronychia: Inflammation and/or infection involving the folds of tissue surrounding the nails.

Pterygium: Destruction of the matrix with subsequent scarring of the proximal nail fold area.

Splinter hemorrhage: A nail change formed by the extravasation of blood from the longitudinally oriented vessels of the nail bed.

Spongiosis: Intracellular edema of the spongy layer of the skin.

Subungual: Beneath the nail

Wickham's striae: Pale, grayish dots or lines forming a network on the surface of the papules characteristic of lichen planus.

REFERENCES

1. Daniel CR III. An approach to initial examination of the nail. In: Scher RK, Daniel CR III, eds. *Nails: Therapy, Diagnosis, Surgery*. Philadelphia: W.B. Saunders Company; 1990;78-81.

2. Basuk PJ, Scher RK, Ricci AR. Dermatologic diseases of the nail unit. In: Scher RK, Daniel CR III, eds. *Nails: Therapy, Diagnosis, Surgery*. Philadelphia: W.B. Saunders Company; 1990;127-152.

3. Rongioletti F, Agostino P, Tripoli S, et al. Proximal white subungual onychomycosis: A sign of immunodeficiency. *J AM Acad Dermatol*. 1994; 30: 129-130.

4. Jerasutus S. Histology and histopathology. In: Scher RK, Daniel CR III, eds. *Nails: Therapy, Diagnosis, Surgery*. Philadelphia: W.B. Saunders Company; 1990; 60.

5. Daniel CR III, Sams M Jr, Scher RK. Nails in systemic disease. In: Scher RK, Daniel CR III, eds. *Nails: Therapy, Diagnosis, Surgery*. Philadelphia: W.B. Saunders Company; 1990:184.

6. Jerasutus S. Histology and histopathology. In: Scher RK, Daniel CR III, eds. *Nails: Therapy, Diagnosis, Surgery*. Philadelphia: W.B. Saunders Company; 1990:63.

7. Zaias N, Ackerman AB. The nail in Darier-White disease. *Arch Dermatol*. 1973;107:193-199.

8. Daniel CR III. Onycholysis: An overview. *Sem Dermatol*. 1991;10:34-40.

9. Daniel CR III, Odom RB. Nail problems from A to Z. *Patient Care*. 1989;23:91-119.

10. Silverman RA. Pediatric disease. In: Scher RK, Daniel CR III, eds. *Nails: Therapy, Diagnosis, Surgery*. Philadelphia: W.B. Saunders Company; 1990;82-105.

11. Daniel CR III, Sams M Jr, Scher RK. Nails in systemic disease. In: Scher RK, Daniel CR III, eds. *Nails: Therapy, Diagnosis, Surgery*. Philadelphia: W.B. Saunders Company; 1990;170.

12. Hudson JB, Dennis AJ Jr. Transverse white lines in the fingernails after acute and chronic renal failure. *Arch Intern Med*. 1976;117:276.

13. Welter A, Michaux M, Blondeel A. Lignes de Mees dans un cas d'intoxication aigue a l'arsenic. *Dermatologica*. 1982;165:482.

14. Mees RA. Eeen verschinjsel by polyneuritis arsenicosa. *Ned Tijdschr Geneeskd*. 1919;1:391.

15. Daniel CR: Unpublished observation. Jackson, Mississippi. Year unknown.

16. Daniel CR III, Bower JD, Daniel CR Jr. The half and half fingernail: The most significant onychopathological indicator of chronic renal failure. *J Miss State Med Assoc*. 1975;6:376.

17. Daniel CR III, Bower JD, Daniel CR Jr. The half and half fingernail, a clue to chronic renal failure. *Proc Clin Dial Transplant Forum*. 1975;5:1.

18. Daniel CR, Osment LS. Nail pigmentation abnormalities, their importance and proper examination. *Cutis*. 1980;25:595.

19. Haley L, Daniel CR III. Fungal infections. In: Scher RK, Daniel DR III, eds. *Nails: Therapy, Diagnosis, Surgery*. Philadelphia: W.B. Saunders Company; 1990;106-119.

20. Zaias N. Onychomycosis. *Arch Dermatol*. 1972;105:263-274.

21. Adapted from Scher R, Daniel CR. *Nails: Therapy, Diagnosis, Surgery*. Philadelphia: W.B. Saunders Company; 1990;123.

22. Baran R. Nail changes in general pathology. In: Pierre M, ed. *The Nail*. New York: Churchill Livingstone; 1981.

23. Daniel CR III. Paronychia. In: Greer KE ed. *Common Problems in Dermatology*. Chicago: Year Book Medical Publishing; 1988; 249-255.

24. Shelly WB, Ralson EL. Paronychia due to an enchondroma. *Arch Dermatol*. 1964;90:412.

25. Herman EW, Kezis JS, Silvers DN. A distinctive variant of pernio. *Arch Dermatol.* 1981;117:26.

26. Kahn G. Nail involvement in histiocytosis X. *Arch Dermatol.* 1969;100:699.

27. Timpatanapong P, Hathirat P, Isarangkura P. Nail involvement in histiocytosis X. *Arch Dermatol.* 1984;120:1052.

28. Baran RJ, Dawber RPR. *Diseases of the Nails and Their Management.* Oxford: Blackwell Scientific Publications; 1984.

29. Stone OJ. Bandage-induced nail disorders. *Cutis.* 1985;36:259.

30. Barrow MV. The nails in multicentric reticulohistiocytosis. *Arch Dermatol.* 1967;95:200.

31. Baumal A, Robinson MJ. Nail bed involvement in pemphigus vulgaris. *Arch Dermatol.* 1973;197:151.

32. Ganor S. Chronic paronychia and psoriasis. *Br J Dermatol.* 1975;92:685.

33. Patterson JW. Pterygium-inversum-unguius-like changes in scleroderma. *Arch Dermatol.* 1977;113:1429.

34. Daniel CR III, Scher RK. Nail changes secondary to systemic drugs or ingestants. *J Am Acad Dermatol.* 1984;10:250.

35. Voorhees JJ, Organos CE. Oral retinoids. *Arch Dermatol.* 1981;117:418.

36. Lovy MR, Bluhm GB, Morales A. The occurrence of nail pitting in Reiter's syndrome. *J Am Acad Dermatol.* 1980;2:66.

37. Selmanowitz VJ, Stiller MJ. Rubinstein-taybi syndrome. *Arch Dermatol.* 1981;117:504.

38. Chanda JJ, Callen JP. Stevens-Johnson syndrome. *Arch Dermatol.* 1978;114:626.

39. Mackie RM. Lupus erythematosus in association with finger clubbing. *Br J Dermatol.* 1973;89:533.

40. Rook A. Naevi and other developmental defects. In: Rook A, Wilkinson DS, Ebling FJG eds. *Textbook of Dermatology.* 3rd ed. Oxford: Blackwell Scientific Publications;1979.

41. Miller SJ. Nutritional deficiency and the skin. *J Am Acad Dermatol.* 1989;21:1.

42. Daniel CR. Paronychia (CME Section). J Am Acad Dermatol. 1994;31:515-518.

43. Baran R. Paronychia. American Academy of Dermatology Annual Meeting, Las Vegas, December 8, 1985.

44. Daniel CR III. Paronychia. In: Greer KE ed. *Common Problems in Dermatology.* Chicago: Year Book Medical Publishing; 1988;249-255.

45. Kaplan MH, Sakick N, McNutt NS, et al. Dermatologic findings and manifestations of acquired immunodeficiency syndrome (AIDS). *J Am Acad Dermatol.* 1987;16:485-506.

46. Dompmartin D, Dompmartin A, Deluoi AM, et al. Onychomycosis and AIDS; Clinical and laboratory findings in 62 patients. *Int J Dermatol.* 1990;29:337-339.

47. Elmets C. Management of common superficial fungal infections in patients with AIDS. *J Am Acad Dermatol.* 1993;31:S60-S63.

48. Elewski BE. Clinical pearl: Proximal white subungual onychomycosis in AIDS. *J Am Acad Dermatol.* 1993;29:631-632.

49. Daniel CR III, Norton LA, Scher RK. The spectrum of nail disease in patients with human immunodeficiency virus infection. *J Am Acad Dermatol.* 1992;27:93-97.

50. Daniel CR III. An approach to initial examination of the nail. In: Scher RK, Daniel CR III, eds. *Nails: Therapy, Diagnosis, Surgery.* Philadelphia: W.B. Saunders Company; 1990;80-81.

APPENDIX A *Nail Questionnaire*[50]

YOUR NAME: _____ AGE: ____ SEX: ____ DATE: _____

1. Were you born with this problem?
   When did you first ever have a problem like this?

2. Which nails were affected first? (Place an * by these)
   Which are affected now? (Place a ✔ by these)

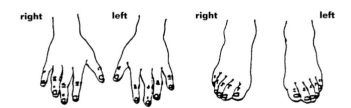

3. How has this changed from beginning to now?

4. Describe your nails in general (hard, brittle, soft, etc.).

5. Have you ever traumatized any of the involved nails (stubbed your toe, hit the fingernail or toenail with a hammer, caught it in a door, etc.)?

6. What kind of work do you do? Do you do anything to affect your nails or the tips of your fingers or toes? Contact with chemicals or irritants, such as strong soaps, hair straightener, lye, etc.? Hands or feet in water a lot?

7. List hobbies in which you might traumatize or otherwise affect your nails (tennis, jogging, painting, playing the piano, etc.).

8. Did you in the past or have you recently
   a. Pick at your nails?
   b. Bite or suck on your nails?
   c. Tear your nails off?
   d. Have ingrown nails?
   e. Wear tight or pointed-toe shoes?
   f. Push the cuticle back (how often?)
   g. Remember "runarounds" or swelling around cuticles?

9. Personal nail care
   a. List any nail cosmetics or conditioners that you use:
      1. Base coat
      2. Top coat
      3. Enamel
      4. Nail strengtheners
      5. Nail hardeners
      6. Cuticle treatment
      7. Gloss
      8. Nail conditioners
      9. Other
      10. Please bring any of these products with you on your next office visit, along with list of ingredients, if possible.
   b. List any instruments that you use to care for your nails.
      What do you do with these instruments?
      How often do you do this?
   c. Do you go to a manicurist? How often?
      What is usually done to your nails?
   d. Have you ever had the following? (If so, how often and when was the last time?)
      1. Sculptured nails
      2. False or artificial nails or "gel" nails
      3. Nail "wraps"
      4. Acrylic nails
      5. Other

10. Do you have any other skin or hair problems, or have you ever had any in the past?
    Lichen planus
    Psoriasis
    Ringworm
    "Jock itch"
    Athlete's foot
    Vaginal yeast infection or other yeast infection
    Other

11. List any medical problems that you have had in the past or have now (diabetes, heart trouble, thyroid problems, etc.).

12. List any medications that you have taken the last year (sulfa, water pills, tetracycline, high-blood-pressure medicines, constipation medicine, chemotherapy, pain pills, vitamins, oral contraceptives, oral pills to help you tan, pills for psoriasis therapy, retinoic acid, NeoSynephrine, etc.).

13. What treatment (self and professional) have you had for your nail problem (past and present)?
    a. List pills and dates used.
    b. List topical treatments and dates used.
    c. List surgical treatment and dates performed.

14. Does anyone in your family have
    a. Nail problems?
    b. Diabetes?
    c. Skin problems (psoriasis, lichen planus, fungus, etc.)?
    d. Thyroid problems?

15. What do you think is the cause of your nail problem?

APPENDIX B *Selected Bibliography*

1. Daniel CR III. Nail Disease in Patients with HIV Infection. Dermatology, Progress and Prospective. The Proceedings of the 18th World Congress of Dermatology. New York: Parthenon Publishing Company; 1993;382-385.

2. Daniel CR III. The Diagnosis of Nail Fungal Infection. *Arch Dermatol.* 1991;127:1566-1567.

3. Daniel CR III. Non-Fungal Infections. In: Scher, RK, Daniel CR III, eds. *Nails: Therapy, Diagnosis and Surgery.* Philadelphia: W.B. Saunders Company; 1990;120-126.

4. Daniel CR III. Onycholysis. Seminars in Dermatology. 1991;10:34-40.

5. Daniel CR III. Paronychia. In: Greer KE, ed. *Common Problems in Dermatology.* Chicago: Year Book Medical Publishing; 1988;249-255.

6. Daniel CR III. The Nail. In: Sams WM, Lynch P, eds. *Principles and Practice of Dermatology.* New York: Churchill Livingstone; 1990;743-760.

7. Daniel CR III. The Nail. In: Daniel CR III, ed. *Dermatologic Clinics.* Philadelphia: W.B. Saunders Company; 1985;Vol 3.

8. Daniel CR III, Norton LA, Scher RK. The Spectrum of Nail Disease in Patients with HIV Infection. *J Am Acad Dermatol.* 1992;27:93-97.

9. Daniel CR III, Scher RK. The Nail. In: Sams WM, Lynch P, eds. *Principles and Practice of Dermatology.* New York: Churchill Livingstone; Second edition;in press.

10. Haley L, Daniel CR III. Fungal Infections of the Nail. In: Scher RK, Daniel CR III, eds. *Nails: Therapy, Diagnosis, Surgery.* Philadelphia: W.B. Saunders Company, 1990;106-119.

11. Scher RK, Daniels CR III, eds. *Nails: Therapy, Diagnosis, Surgery.* Philadelphia: W.B. Saunders Company; 1990.

12. Elewski, BE. Clinical Pearl: Diagnosis of onychomycosis. *J Am Acad Dermatol.* 1995;32:500-501.